ALLE ZEIT WACH 1842

R. Steinberg (Ed.)

Music and the Mind Machine

The Psychophysiology and Psychopathology of the Sense of Music

Springer-Verlag
Berlin Heidelberg New York
London Paris Tokyo
Hong Kong Barcelona
Budapest

Professor Dr. REINHARD STEINBERG
Pfalzklinik Landeck
Weinstraße 100

D-76889 Klingenmünster
Germany

ISBN-13: 978-3-540-58528-2 e-ISBN-13: 978-3-642-79327-1
DOI: 10.1007/978-3-642-79327-1

Typesetting: Camera ready by author
25/3130-5 4 3 2 1 0 – Printed on acid-free paper

Preface

Two Tchaikovsky enthusiasts, Mr. A and Mr. B, listen to a performance of Tchaikovsky`s 5th symphony. They are sitting next to each other and have the best seats in the concert hall. Both men listen intently and with great interest and hear exactly the same music, also the same room acoustics, loudness, timbre, and so on. At the end of the piece Mr. A leaps to his feet and applauds frenetically: He found this music and its performance deeply satisfying. He even comments to his wife that in spite of everything there are good things in this world, that this music is divine. Mr. B did not clap; he even booed twice and left the concert hall infuriated with this performance. He told his wife that the conductor and the orchestra lacked in everything this music needed, that the mendacity of the conductor carried over to the orchestra, which is the victim of his ignorance, and so on.

What happened to Mr. A and Mr. B. They both went to the concert in anticipation of a great musical event. Admittedly, such contrary reactions will be rare with Tchaikovsky, but we find an increasing number of Mr. As and Mr. Bs in reactions to modern music or to new productions of operas, for instance. Just remember the enormous scandal at the first performance of Strawinsky`s "Rites of the Spring" in 1913 in Paris or the Chereau/Boulez production of Wagners "Ring des Nibelungen" in the late 1970s in Bayreuth!

But novelty is not the problem with accepting Tchaikovsky and the special enthusiasm both gentlemen feel for this composer. Maybe they were victims of their anticipations, of which they were only partly aware, and which like a self-fulfilling prophecy confirmed their positive or negative prejudice against the performers. Psychologists call it prestige suggestion, a mechanism which not only applies in the field of music. But this explanation is probably too simple and does injustice to both enthusiasts. Such injustice can only be avoided by questioning the men and trying to understand what they expected, what they understood, what in general they feel when they hear music, above all when they hear their favorite composition, how much music education they have had and so on.

Man has been making music since the beginning of time and since then he has been developing theories about its origin, its importance, its meaning, its representation in our mind, and its correct or incorrect performance. In former centuries music was believed to to be a gift of God. Its theory was part of all studies. It was taught in medicine as well as in mathematics. Today music is omnipresent part in our lives as never before. In sharp contrast, our theories about music, its meaning, and its origin remain poor.

What could Mr. B discuss with Mr. A about this concert; how would they argue about the performance and their contradictory impressions? They both could claim that unwritten strict rules are taught at the music academies, which apply to a proper performance. It wouldn't get them very far, though, since these rules did not really seem to be violated. Could science perhaps help them to discuss the matter more productively?

Natural science has been meeting with music unobtrusively for about a hundred years, almost overlooked by the main stream of scientific efforts. Nevertheless within the last three decades very different branches of science have begun defining mechanisms of the mind's workings, which are the basics of our understanding of music. People who conduct this research have only their interest in music in common. Multidisciplinarity characterizes this small group of researchers; they come from the fields of music psychology, musicology, neurology, physics, mathematics, developmental psychology, psychiatry, neurophysiology, art therapy, psycho-acoustics, neuroanatomy, and yet others.

In September 1992 a conference was held in Munich to present the state of the art in the research of the musical mind. Acknowledged experts in their fields reported on new results. Making music integrates motor skills, acquired musical techniques, training and knowledge as well as emotional contents and cognitive strategies. Music has the advantage over language of being well definable in physical terms, without the intervening property of semantic meaning. Loudness, timbre, pitch, harmony, rhythm, and many other features characterize music qualitatively and quantitatively, allowing for an experimental approach which focuses on simple variables. This positivistic scientific approach requires, of course, reducing music to some basic elements. Nevertheless, findings in neuroscience can be correlated with musical elements. This kind of approach has given us much insight into cognitive mechanisms, even in neuronal representations of the sense of music. Experiments with trained musicians and people with no music education at all have been very important. Careful studies in patients after a cerebral stroke which resulted in moderate or even complete amusic disturbances are as helpful as studies about the change within the musical sense of mental patients. Both models offer the chance to follow up the natural course of regeneration of both generally disturbed cognitive mechanisms and those necessary for the musical mind, which in turn helps us to understand normal physiology. The new techniques of brain activity imaging are very helpful in this kind of research.

Music perception and music understanding otherwise require wholistic approaches which are mainly accessible by using music psychological strategies. This includes learning music and becoming acquainted with the systematology of a musical language. The process starts at birth and, in principle, never ends. How do we learn to visually distinguish a figure from a background. In musical terms, to recognize a meaningful musical structure out of other auditory events. Why does a piece by Scott Joplin give rise to other emotions than music by Johann Sebastian Bach or a Vienna waltz which conjures up an image of dancing in many people? What about the therapeutic effect of music? In scientific terms we are almost totally lacking insight into those mechanisms, although there is no reason to doubt this very important effect of music. Music seems to give us access to cerebral stroke patients, which has been overlooked for no reason for a long time.

The conference participants could have discussed the very contradictory musical adventure with Mr.A and Mr.B. Hypothetically it could have helped Mr. B to overcome his anger. He could have learned that, according to the critics in his favorite newspaper, from a scientific point of view the performance was quite good technically and that he wasn't well that evening. As for Mr. A, there would be a real risk of disillusionment if he were to read in the newspaper the next day that the performance was only mediocre as regards the standard of the orchestra and the conductor. At the next concert both would understand a bit more about some of the principles of the musical mind but the fact remains that they will still have strong and even contradictory emotional reactions and opinions. It is also certain that a scientific explanation of elementary rules of the musical mind will never create a composition or even help a composer to write a piece of music which is more accepted by his audience.

For science as well as the arts pleasant surroundings are stimulating and inspiring. The beautiful location and the hospitality of the Carl Friedrich von Siemens Stiftung is most gratefully acknowledged, as is the grant from the Deutsche Forschungsgemeinschaft. I wish to express my deepest gratitude to Hanns Hippius, Director of the Psychiatric Hospital of the University of Munich, who supported the organization of the conference in very many ways, especially through the support of the "Münchener Verein zur Förderung der Psychiatrie e.V. " I am also indebted to many companies for substantial help in bringing together researchers from all over the world. I gratefully acknowledge the competent help of Springer-Verlag, especially the stimulating commitment of Sherryl Sundell. Finally I thank my coworkers Robert Rohmfeld, Markus Fani, and, especially, Dorothea Dietschmann, who made even impossible things happen.

Reinhard Steinberg
October 1994

Psychiatric University Hospital, Munich
Pfalzklinik Landeck, Klingenmünster

Contents

I. Musical Expression and Comprehension

II. Music Physiology and Pathophysiology

List of Contributors

Beisteiner, Roland, Dr. med.
Neurologische Klinik der Universität, Währinger Gürtel 18-20, A-1090 Wien, Austria

Bharucha, Jamshed J., Ph D
Dartmouth College, Dept. of Psychology, Hanover, NH 03755, USA

Carterette, Edward C., Ph D
Departments of Psychology and Ethnomusicology & Systematic Musicology, University of California, Los Angeles, California 90024, USA

Davidson, Jane W., Dr.
City University, Northampton Square, London EC 1V OHB, UK

Deecke, Lüder, Prof. Dr. med.
Neurologische Universitätsklinik, Währinger Gürtel 8, A-1090 Wien, Austria

Deliège, Irène, Ph D
Research Unit in the Psychology of Music, University of Liège,
5 Boulevard du Rectorat, b 32, Start Tilman, B-4000 Liège, Belgium

Ebner-Pfeuti, Rosemarie
Bündnerstr.20, CH 4050 Basel, Switzerland

Gabrielsson, Alf, Ph D, Prof.
Uppsala University, Box 1854, S-75148 Uppsala, Sweden

Hassler, Marianne, Priv.-Doz., Dr. rer. soz.
Psychologisches Institut der Universität, Abtlg. für klinische und physiologische Psychologie, Hauserstr.114, D-72076 Tübingen, Germany

Mazzola, Guerino, Priv.-Doz., Dr. rer. nat.
Mathematisches Institut der Universität Zürich, Wangenstr.11, CH-8600 Dübendorf, Switzerland

Motte-Haber de la, Helga, Prof. Dr. phil.
Technische Universität, Fachbereich 1, Fachgebiet Musikwissenschaft,
Sekr. H 63, Straße des 17. Juni 135, D-10623 Berlin, Germany

Nielzén, Sören, M D
University Hospital, Dept. of Psychiatry, Experimental Music Research,
S-221 85 Lund, Sweden

Papousek, Hanus, Prof. Dr. med.
Papousek, Mechthild, Priv.-Doz. Dr. med.
Max-Planck Institut für Psychiatrie, Kraepelinstr. 10, D-80804 München, Germany

Peretz, Isabelle, Ph D
Département de Psychologie, Université de Montreal, CP 6128 succ. A,
Montreal (Que), H3C 3J7, Canada

Petsche, Hellmuth, Prof. Dr. med.
Neurophysiologisches Institut der Universität, Institut für Hirnforschung der Österreichischen Akademie der Wissenschaften, Währingerstr. 17, A-1090 Wien, Austria

Pöldinger, Walter, Prof. Dr. med.
Psychiatrische Universitätsklinik, Wilhelm Klein Str. 27, CH-4025 Basel, Switzerland

Raith, Lydia, Dr. med.
Psychiatrische Klinik der Universität, Nußbaumstr. 7, D-80336 München, Germany

Richter, Peter, Dr. med.
Neurophysiologisches Institut der Universität, Währinger Str.17, A-1090 Wien, Austria

Steinberg, Reinhard, Prof. Dr. med.
Psychiatrische Klinik der Universität, Nußbaumstr. 7, D-80336 München, Germany
Pfalzklinik Landeck, Weinstraße 100, D-76889 Klingenmünster, Germany

Sundberg, Johan, Prof., Ph D
Royale Institute of Technology, Dept. of Speech Communication and Music Acoustics, S-100 44 Stockholm, Sweden

Terhardt, Ernst, Prof. Dr. rer. nat.
Technische Universität, Lehrstuhl für Elektroakustik, Arcisstr.21,
D-80333 München, Germany

Wieser, Heinz-Gregor, Prof. Dr. phil.
Neurologische Klinik der Universität, Frauenklinikstr. 26, CH-8091 Zürich, Switzerland

I. Musical Expression and Comprehension

Convergent Research Methods in Music Cognition *

EDWARD C. CARTERETTE AND ROGER A. KENDALL

Convergent Research Methods in Music Cognition

An often overlooked principle of perceptual research is that what we learn is dependent upon what we ask and how we ask it. Experimentation in cognitive music research is driven implicitly and explicitly by assumptions, models, and theoretical constructs both of cognitive psychology and music. Most research in music perception and cognition is implemented by a single method. Recently, questions have been raised about the validity and generality of such a limited approach, particularly regarding such complicated theoretical constructs as musical tonality. We propose that a useful technique is to converge on the answer to experimental questions by applying multiple methods, in essence to investigate simultaneously the central research question as well as ancillary questions of method. We suggest that psychomusical research mandates the use of multiple, convergent methods and the invention or derivation of new approaches. This principle of convergence is illustrated by examples from research on tonality, musical communication and timbre.

It is particularly true of psychomusical experiments that a major struggle in research design exists between reliability and validity. The nearer to the natural setting of a live concert, the more musically valid is the empirical context. But variability of this context across time, space, and performances may yield potentially unreliable data owing to many unidentified and uncontrolled variables. Conversely, as the musical universe is limited to single, synthetic sounds, reliability and repeatability are won at the cost of musical validity. Much past psychomusical research has yielded very reliable results, but results of little musical validity or interest. There now exists an extraordinary force and tool of great power which permits a frontal assault on musical validity - the computer.

A computer of astonishing swiftness and gargantuan memory costs very little. There are ingenious and flexible programs for generating and manipulating musical stimuli so as to achieve increasing ecological validity among several dimensions. For example:

1. Synthetic stimuli may be endowed with time-variant properties of very high

* This chapter is based on the ideas and materials of Kendall and Carterette (1992)

complexity. The low cost of hardware and software permits the generation of precise stimuli which can be sounded instantly in any order.

2. Natural signals can be sampled, stored, transformed and controlled at musical-bandwidth rates for studying performer-initiated events, orchestration and related acoustical effects.
3. Flexible programs permit easy control of all classes of instruments and sounds - natural, synthetic, sampled, MIDI - and these may be mixed for contrast and comparison at different levels of musical validity.

Although this great computer power is easy to come by, researchers use a limited repertory of methods. Why? Besides fear, ignorance and complacency, there are many reasons:

1. Some computer literacy and knowledge of signal properties are required.
2. Hardware and software systems for experimental control are not integrated or interfaced with mathematical and statistical programs for analysis of signals and response data. This begets ponderous, repetitive processing.
3. Control systems often are built around a single method. In order to use a new paradigm, a new systems program must be written, which takes time and money. Thus researchers may be compelled to tailor their research to the available resources.

A sad result is that research questions often are driven by the method at hand rather the method being driven by the question, or even invented to serve the question.

Another limit of current psychomusical research is set by the reification of concepts. Concept reifiers presume that objects and concepts named in a theory, such as tonality, pitch, timbre, loudness, or rhythm, are orthogonal to each other in their unique, invariant properties. But in truth the researcher utilizes an explicit or implicit model, which entails a set of interrelated relationships among variables. The variables and the hypothetical relations among them are established by the researcher. Models are relative and are observer based. We create the experimental world in images of our own theories. The value of a model is based on the descriptive, organizational, and predictive relations between variables set by the theory and scientifically tested. Experimental methods and designs must take into account both reliability and validity. No single model will completely account for all the observables of the real world even in the case of a single concept, such as tonality or pitch or timbre.

Recently, we have suggested (Kendall and Carterette 1990) that psychomusical research must abandon the privileged status accorded to single frames of reference such as the notational and acoustical and must include aspects of the complex interrelations among composer, performer, and listener as the musical message is recoded to and from the notational and acoustical signal by these three behavioral systems. Thus a given musical concept can have a different structure and definition at many levels of musical communication. The methodological implications of such a position are profound. Operational definitions of concepts identified in theory are reflected directly in the methods and techniques. Different methods can and often do reflect different facets of a concept. We argue that the myopic focus on singular methods often obscures the multidimensional nature of musical communication.

The Principle of Convergence

Convergence can be put in binary terms of the necessary [Nec] and sufficient [Suf] conditions for obtaining a given result. The four states are: <Nec, Suf>, <Nec, not Suf>, <not Nec, Suf>, and <not Nec, not Suf>. Let us take an elementary, theoretical case: An experiment uses two methods, A and B, differing in a single attribute X such that A lacks X and B possesses X. If both methods A and B yield equivalent results, then we have X:<not Nec,Suf> because A lacking X does as well as B possessing X. The matrix below shows all possible cases.

Outcome interpretable	Nec, Suf conditions for X
A yes B yes	< not Nec, Suf >
A yes B no	< not Nec, not Suf >
A no B yes	< Nec, not Suf >
A no B no	< Nec, not Suf >

I will not dwell on this logic but simply say that Kendall and I reason this way in, for example, our studies of verbal attributes of timbre, which I discuss shortly. A fuller discussion of the logic is given by Kendall (1986).

In the real world the differences between methods may be multivariate or the two methods A and B may both yield interpretable results which differ in meaningful ways, thus giving evidence that a multidimensional approach to the experimental question is required. In this case we still converge on different facets of the truth, and additional, collateral studies using yet different methods are required.

Methods can be varied in both stimulus and response properties and in temporal relations between them. Convergence on a multidimensional concept often requires varying aspects of both stimulus and response and translating subjects' data into comparable formats. For example, ratings can be changed into similarities and vice versa. In illustration of these ideas, let us turn to some experiments which vary method.

Tonality

A lively controversy over tonality recently has been aired (Krumhansl, 1990a; Butler, 1990). Tonality is said to be "loyalty to tonic" or "a tendency for 'active' tones in a melody to 'move' toward 'rest'." Imagining a plagal cadence of IV-I (A-men) at the end of an Anglican hymn almost induces a kinesthetic feeling that the harmony on "A" must resolve to the chord on "men." The Gestalt principle of closure is closely related to the concept of tonality in melodic and harmonic motion.

Krumhansl's (1990a) "tonal hierarchy" is defined by an operation:

> In the basic experiments, the elements of interest (tones or chords) are presented singly or in pairs following a musical unit, such as a scale or chord cadence....The listeners are required to judge, using a numerical scale, how related the presented elements are....The rationale for presenting tones in key-defining contexts is that key, or tonality....established hierarchies of structural significance for tones and chords (p. 10).

This "probe-tone" method has been at the heart of a very large number of studies. In fact, this method measures the degree of closure. The scale or chord progression is the context and the rating is an estimate of the degree to which the "elements of interest" close the pattern. The controversy is: Does the probe-tone method measure "tonality"?

Butler (1989, 1990) contends that the probe-tone rating is unidimensional but is applied to a stimulus which is multidimensional. The temporal ordering of tones strongly influences these ratings. Thus the closure aspect of the probe-tone rating leads to invalid results. Krumhansl's method deals with zero-order Markov chains. Butler's remedy is based on intervals and so is first-order, modeled on the principle that "any tone will suffice as a perceptual anchor - a tonal center - until it is supplanted by a better rival" (Butler 1989, p. 429). Though his experiments, with probe-tones on time-reordered pitch sets, make trouble for Krumhansl's method, he offers no substitute. Alas, his suggested response method "...vocalizing, whistling, or by using sol feggio syllables..." appears to be open to experimenter interpretation.

Treating tonality as orthogonal to time especially using zero-order chains considerably limits musical validity. Nonetheless Krumhansl's probe-tone results produce profiles of pitch-class relations which are modestly confirmed by information-theoretical analysis of notation (Krumhansl 1990, p. 157). Krumhansl's work includes probe-tone variants, and she admits the role of time in her concept of contextual asymmetry (Krumhansl 1990b, p. 149). And the probe-tone method confirms much of her earlier work which uses similarity scaling (Krumhansl 1990, p. 148).

We believe that the use of probe-tones, zero and first order chains, and similarity scaling - all these methods - illuminate different aspects of "tonality." But unless multiple methods are conducted at the same time in the same experimental context, it will be hard to tell which procedures are providing the greatest ecological validity. Such factors as rhythm, contour, and timbre must be involved if we are to discover how tonality and other musical materials interact. As an example, it should be possible to alter the tonal implications of a sequence of pitch classes by varying the accenting of differing groups. Recently, a graduate student in our laboratory, Michael Frischkopf, applied similarity scaling to paired patterns of differing Markov-chain orders, with differing frequency distributions of pitch classes. Similarity scaling of patterns which vary in pitch, time, and tonal materials would usefully extend previous work. Such patterns can be categorized, rated for semantic attributes, or used in matching procedures. In short, it may not be necessary to use "multidimensional response measures" as Butler implies. Instead, we can converge on the truth about tonality by means of multiple experiments and methods. These would vary stimulus materials such as chain order, timbre, pitch-time contour, and responses such as probe-tone, similarity, rating and identification. We turn now to work which has applied such multiple methods.

Musical Communication

A good example of the use of several, convergent methods comes from work which involved the frames of reference of the composer, performer, and listener, as well as the notational and acoustical frames.

> We conceive of music as based on a system of sequences of state that fill gaps of inexorable time. Depending on the frame of reference, these states can be, for example sound states (perception), acoustic states (vibrational signal), symbol states (notational signal), and mental states (cognitive metasymbols). In fact, the transformations from one state sequence to another is the very core of the difficulty in musical communication. (Kendall and Carterette 1990, p. 132).

In studying the musical communication model, we asked a professional pianist to scan a phrase from "Thy Hand, Belinda" by Purcell (Fig. 1) and to produce three expressive performances: Without expression (senza espressione), with appropriate

Fig. 1. Notation of excerpt from "Thy Hand Belinda" by Purcell. The *bracketed phrase* was the musical material of the experiments. (From Kendall and Carterette 1990)

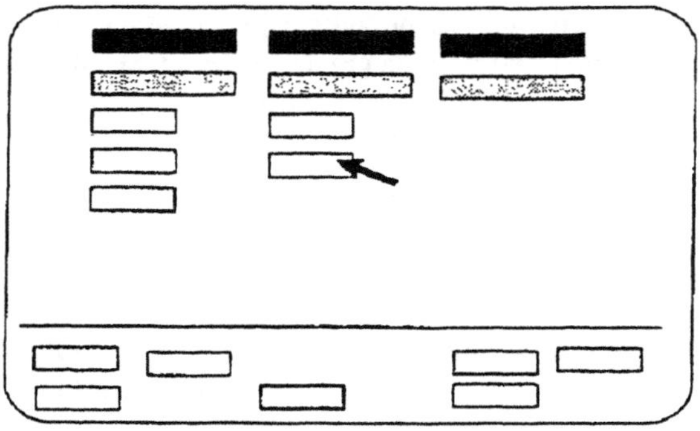

Fig. 2. Schematic diagram of the computer screen seen by a subject during categorizing. (From Kendall and Carterette 1990)

expression (con espressione), and with exaggerated expression (con troppo espressione). Four instrumentalists - oboist, violinist, clarinetist, and trumpet players - listened to the piano models and performed their own renditions of the three expressive messages.

In assessing the degree to which the intended level of expression was communicated to the listener, several methods were used. The first was a computer-controlled categorization task: A set of three yellow rectangles, representing the model stimuli, were shown on the computer screen (Fig. 2). The set of 12 choices, four wind-instruments x three expressive levels, were represented by uniform white rectangles at the bottom of the screen. All stimuli were randomly ordered. The subject could hear any stimulus by pointing at the rectangle and clicking a mouse button, then could move a choice in position below one of the three expressive models. Choices could be moved freely and any choice or model could be heard as often as desired. The program recorded the final configuration as well as the number of category changes, number of auditions, and listening durations.

Confusion matrix data from 19 subjects, 10 nonmusicians and nine musicians, showed correct matches ranging from 81% (musicians, clarinet) to 43% (nonmusicians, violin) (Table 1). A composite matrix summed over instrument showed 70% correct categorizations for the musicians and 64% for the nonmusicians (Table 2). Listening durations, number of times auditioned, and number of category changes indicated that there was a higher cognitive load in discriminating between con espressivo and con troppo espressivo, than between senza expressivo and the other expressive levels (Table 3).

The next method employed was matching, wherein the original stimuli were used with a different response paradigm. The task was a three-alternative forced-choice with free review. On each trial, one of the 12 choice stimuli was heard, followed by a random ordering of all three piano renditions. The subject chose the piano rendition that best fit the level of expressiveness of the initial example. The subject could replay all stimuli. Results from 10 musicians showed correct responses ranging from 55% to 58% (Table 4).

In the categorization paradigm, a subject can review any or all stimuli. Thus the listener can build a perceptual or cognitive space. In contrast, matching restricts comparisons between categories and emphasizes the current model and its choice

Table 1. Categorization frequencies

Expressive level	Clarinet P1	Clarinet P2	Clarinet P3	Oboe P1	Oboe P2	Oboe P3	Violin P1	Violin P2	Violin P3	Trumpet P1	Trumpet P2	Trumpet P3
Musicans												
1	8	1	0	8	1	0	9	0	0	8	1	0
2	1	6	2	1	6	2	0	2	7	1	6	2
3	0	1	8	0	3	6	0	7	2	0	2	7
Mean hits (%)			81			74			48			78
Nonmusicians												
1	6	3	1	6	2	2	8	1	1	8	0	2
2	1	7	2	1	6	3	1	2	7	2	6	2
3	2	2	6	2	2	6	1	6	3	2	3	5
Mean hits (%)			63			60			43			70

P, performance

Table 2. Categorization frequencies summed over instruments

Expressive level	Musicians P1	Musicians P2	Musicians P3	Nonmusicians P1	Nonmusicians P2	Nonmusicians P3
1	33	3	0	28	5	7
2	3	20	13	6	21	13
3	0	13	23	6	14	20
Mean hits (%)			70			64

P, performance

set. In terms of necessary and sufficient conditions, between-category comparison is necessary and sufficient to achieve relatively high hit rates (ca. 70%+). However, since both methods produced the same profile of hits which were above chance (33%), both methods showed that performer intent was communicated.

Table 3. Mean listening durations

Expressive level	Clarinet	Oboe	Violin	Trumpet
1	49,74	53,11	36,37	45.05
2	73,05	68,90	77,32	59,37
3	71,00	67,63	70,37	57,90

Table 4. Match frequencies (musicians)

	Clarinet			Oboe			Violin			Trumpet		
Expressive level	P1	P2	P3	P1	P2	P3	P1	P2	P3	P1	P2	P3
1	11	7	2	10	6	4	14	2	4	14	2	4
2	4	14	2	6	13	1	3	10	7	7	6	7
3	2	9	9	3	5	12	1	10	9	5	2	13
Mean hits (%)			57			58			55			55

P, performance

The next convergent method employed involved a change in stimulus, but used the categorization paradigm. All 15 performances (piano + four wind instruments) x three espressivities were analyzed acoustically in 100-sample (3.5-ms) windows, which yielded both timing and root - mean -square (RMS) amplitude data. These timings were used to drive a MIDI synthesizer which emulated the original instruments, except that amplitudes were fixed, that is, no longer varied in time. Twelve nonmusicians categorized the "synthetic" stimuli comprising the model and choice set. Mean correct matches ranged from 33% (violin) to 69% (clarinet) (Table 5). Yet the same profile of responses by instrument category were found as with the first method which used the natural signals. Thus, we found that timings alone carried enough of the expressive message to be sufficient for reasonable accuracy of categorizing.

The final method employed was rating. Both natural and artificial stimuli were presented randomly and rated by eight musicians who saw on the computer screen a scale running from 0 (without) to 100 (great) expressiveness. Like matching, the rating method restricts intercategory review, and the results (Figs. 3,4) "produced data parallel to those of matching" (Kendall and Carterette 1990, p. 159). Hence

Table 5. Nonmusician categorization frequencies: artificial performances

	Clarinet			Oboe			Violin			Trumpet		
Expressive level	**P1**	**P2**	**P3**	**P1**	**P2**	**P3**	**P1**	**P2**	**P3**	**P1**	**P2**	**P3**
1	7	3	2	7	0	5	5	5	2	8	2	2
2	4	8	0	2	8	2	2	2	8	2	6	4
3	1	1	10	3	4	5	5	5	2	2	4	6
Mean hits (%)			69			56			33			56

P, performance

the method was neither necessary nor sufficient to produce the relatively high accuracy of categorization; it was, however, sufficient to differentiate expressiveness above chance levels for both natural and artificial performances (pp. 157 - 160). Also, the change of response method to ratings, using the verbal attribute expressive, suggested that the "appropriate" level of expression had higher expressivity than the "exaggerated" level (p. 159), though high variability precluded statistical significance. This shows the value of collateral methods not only in confirming previous results, but in suggesting the course for additional research, namely, experiments with the use of rating scales. We turn to a series of experiments which end in the exploration of this issue.

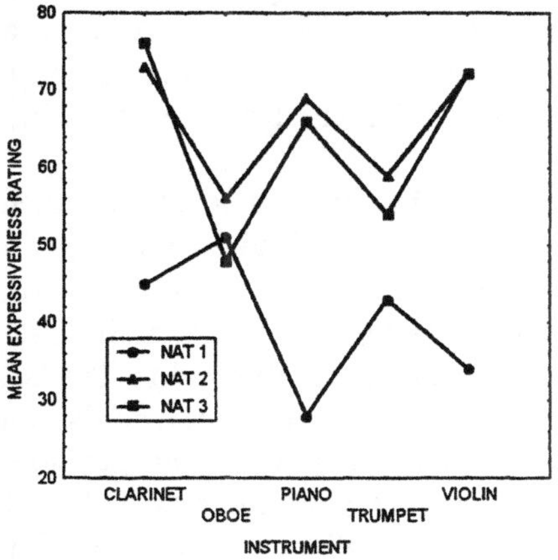

Fig. 3. Mean expressiveness ratings across instruments and levels of expression for natural stimuli (*NAT*). (From Kendall and Carterette 1990)

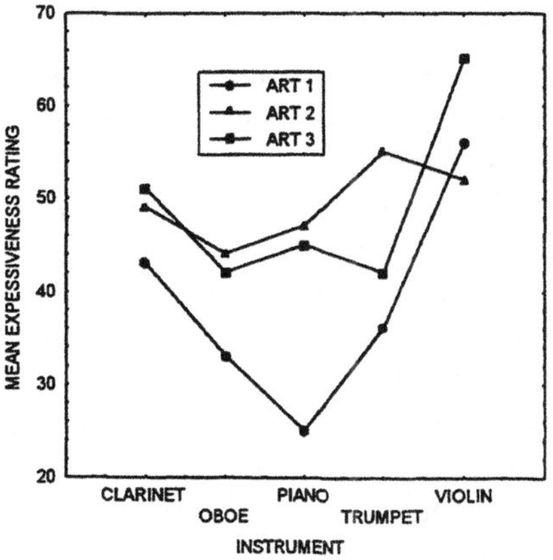

Fig. 4. Mean expressiveness ratings across instruments and levels of expression for artificial stimuli *(ART)*. (From Kendall and Carterette 1990)

Timbre

Kendall and Carterette (1991, 1993a-c) in extensive studies used multiple, convergent methods on the perceptual, verbal, and acoustical attributes of two wind instruments sounding together. In the first method, timbral-similarity ratings were obtained among all pairs of saxophone, oboe, flute, trumpet, and clarinet dyads (thus OC = a dyad of oboe and clarinet). Musical contexts were unisons (Bb4), unison melodies (D5, Eb5, F5, and D5), major thirds (Bb4-D5), and harmonized melodies (using I, IV_6, V_6, I harmonic progression). In the homophonic, harmonized conditions, each member of the dyad appeared as the soprano (top), resulting in six musical contexts. Music majors (n = 19) and nonmusic majors (n = 16) rated similarities of all possible pairs of dyads (n = 10 or a total of 55 items for each context). A 100-point scale appeared on the computer screen. Using a mouse, subject moved a red pointer to the desired position on the scale, then pressed the mouse button. Triangles of dissimilarity ratings were averaged over subjects, then subjected to classical multidimensional scaling (MDS) using Kruskal's method of minimizing stress.

The stress of the resultant three-dimensional configurations ranged from a high of 14.8% for inverted instruments on the harmonized melody for nonmusicians, to a low of 2.0% for major-thirds rated by musicians. Two of the dimensions were stable and interpretable, with associated extremas thus: Dimension 1: oboe (nasal) versus trumpet and saxophone; Dimension 2: saxophone (reedy, rich) versus trumpet (brilliant). The third dimension was interpreted as "simple" versus "complex."

The average dissimilarity triangles were treated as "subjects" in an analysis by individual-differences scaling (INDSCAL). The resulting two-dimensional configuration (Fig. 5) was a circumplex, with extrema as above.

We noted (Kendall and Carterette 1991) that "one weakness of MDS generally is that, once the scaling is complete, the number and 'meaning' of dimensions are subject to interpretation...[this] often has been lax..." (p. 132). Additional experiments, using different methods, were conducted so as to converge on the "meaning" of the dimensions as well as the fundamental nature of timbral combination.

For example, one question which arose is whether the individual instrument timbres linearly sum to form the composite timbre of the dyad combination. To answer this, a "mental imaging" method was devised. The subject, an expert, was given a space marked with the dimensional interpretation (nasal, rich vs brilliant) suggested by the composite MDS two-dimensional (2-D) configuration. The task of the expert musician subject was to imagine the timbre of the individual instruments (clarinet, oboe, saxophone, flute, trumpet) and mark their positions, relative to one another, in a space in which the horizontal axis represented "nasal" versus "not nasal" and the vertical axis represented "brilliant" versus "rich" (Fig. 6). Each of the five instrument points was treated like a vector in 2-D space, such that the location of the combination of instruments in the same space was the mean, unweighted vector sum of the coordinates of each of the constituents of the dyad. This dyad space, which was obtained mathematically from the imaging

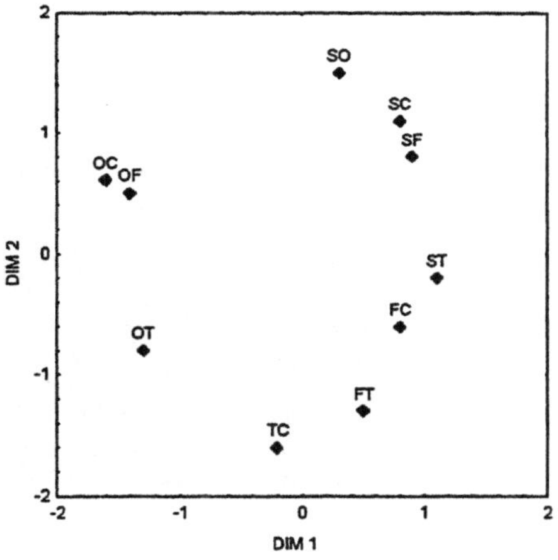

Fig. 5. Two-dimensional (*DIM 1, DIM 2*) INDSCAL solution summarizing wind instrument dyad distances across contexts. DIM 1, *left to right* is "nasal" to "not nasal"; DIM 2, *top to bottom*, is "rich" to "brilliant." *S*, saxophone; *O*, oboe; *C*, clarinet; *F*, flute; *T*, trumpet. (From Kendall and Carterette 1991)

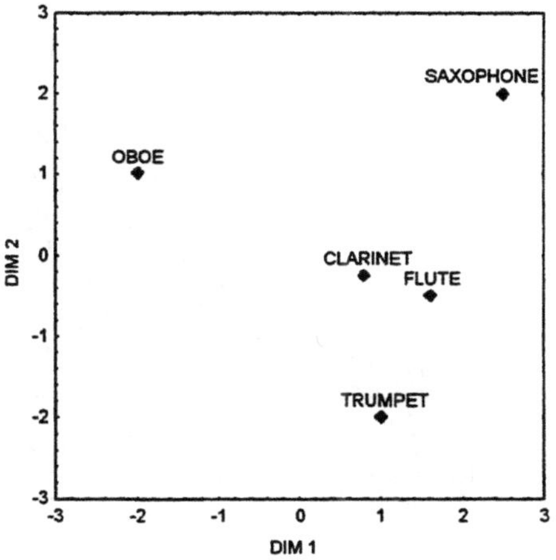

Fig. 6. Two-dimensional configuration of five single instruments positioned by a professor of musicology on the basis of mental image of timbre (Bb4). Dimension 1, *left to right*, is "nasal" to "not nasal"; Dimension 2, *top to bottom*, is "rich" to "brilliant". (From Kendall and Carterette 1991)

experiment, could be compared to the space which was based on dyad timbral similarity: Pearson correlation between distances of the first dimension was .938, between distances of the second dimension was .983 (Kendall and Carterette 1991, p. 396). By use of this "invented" method, initial evidence was obtained for linear combining of timbres and for the validity of the verbal descriptions of the scaling dimensions.

With the next method, we explored additional aspects of timbral combination, specifically the nature of the timbral "fusion" (Kendall and Carterette 1993a). Degree of blend of two instruments was made operational by the extent of "oneness" versus "twoness." Those same dyads of previous experiments were played in random order for each of nine musicians whose task was to move a red pointer on a 100-point scale to the position which best described her feeling of "oneness" or "twoness." Results (Fig. 7) indicated that blend ratings separated instruments along the "nasal" versus "not nasal" axis. "Nasal" combinations blended less well than those which were "brilliant" or "rich."

An additional method was used, aimed at determining a subject's ability explicitly to identify the two instruments sounding. Eight musicians heard randomized soundings of the 55 dyads in six blocks of musical context. The subject then selected, from a list of ten shown on the computer screen, the correct pair of instruments then sounding. As might be expected, an inverse relationship was found between identification (Fig. 8) and blend (Fig. 7): The greater the blend, the worse the identification (for unisons, Pearson $r = -.731$).

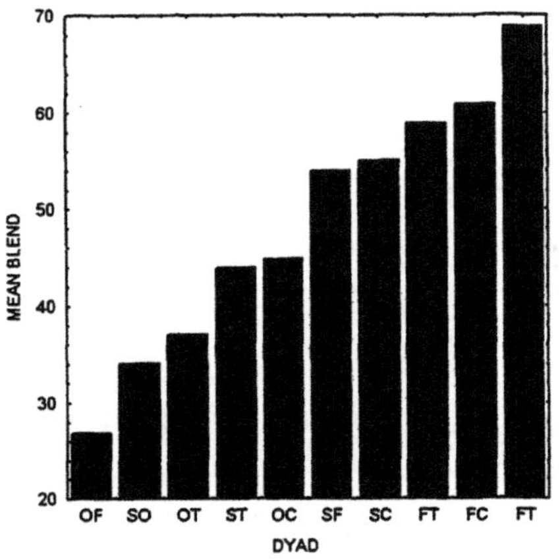

Fig. 7. Rank order, from least to most, of mean blend (averaged over context) for the ten dyads. For abbreviations see legend to Fig. 5. (From Kendall and Carterette 1993a)

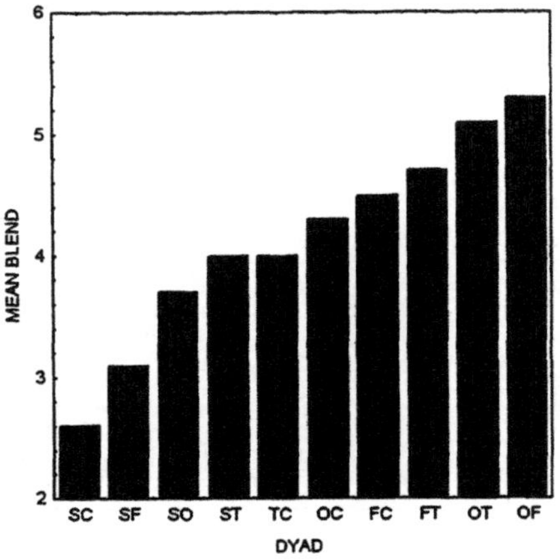

Fig. 8. Rank order, from least to most, of mean correct identification (averaged over context) for the ten dyads. For abbreviations see legend to Fig. 5. (From Kendall and Carterette 1993a)

Next, blend ratings were decomposed into a triangle in which the blend rating for a dyad, OF for example, was an entry in row O column F. This triangular matrix was analyzed by Kruskal's MDS. The result was a 2-D space (stress < .00031) for the five individual instruments. By means of vector addition (see above), these individual points were combined to produce a hypothetical dyad space based on blend. This blend space was then correlated with the composite similarity space, yielding .893 for dimension 1 unisons, and .843 for dimension 2 unisons, showing "a remarkable convergence between differing methods and models" (Kendall and Carterette 1993a).

All these results show how rating, identification, imaging, matching, and similarity-scaling methods converge on the truth in the study of timbral combinations. The next set of experiments required analysis of a single method, the semantic differential. The experiments had a particular purpose: to assign specific appropriate labels to the dimensional axes of the timbral similarity-scaling space.

We used traditional semantic differential scaling on differentials used early by von Bismarck (1974a, b), which we selected for their high factor loadings (Kendall and Carterette 1993a). In order to rate the ten dyad unison timbres of our previous studies, we used eight "bipolar" adjectives: hard - soft, sharp - dull, complex - simple, compact - scattered, pure - mixed, dim - brilliant, heavy - light. The subject judged the verbal attributed of the timbre by using the mouse to position a pointer on a 100-point scale. The eight scales were presented and ratings collected before the next dyad was played. This method differs from paper-based procedures wherein the subject has all of the scales and responses before his eyes. In the computer-based procedure, each scale is presented randomly then erased from the screen after the response.

Results clearly demonstrated the failure of this method to discriminate among the ten unison dyad timbres. We said, "it has occurred to us that a central problem in the use of bipolar opposites is that the 'opposite' is, in fact, not an antipode" (Kendall and Carterette 1993a, p.452). We changed to another method, which we called Verbal Attribute Magnitude Estimation (VAME). In this method, an adjective is paired with its negation, e.g., "sharp" versus "not sharp." The same procedure was used except that the timbre was sounded before every scale. Thus subjects rated the degree of "sharpness," not the degree of "sharp" versus "dull."

Results ($n = 11$) showed: "a far greater degree of differentiation of ratings by instrument..." (Kendall and Carterette 1993b), which was supported by statistical significance for the dyad-by-scale interaction in the analysis of variance (ANOVA) for the VAME method and the lack of the same for the semantic-differential method. Thus, the method of semantic - differential was neither necessary nor sufficient, but the VAME method was both necessary and sufficient to yield interpretable results.

Many various transformations of the ratings data, including assigning to a dyad a vector of attributes and then applying MDS, yielded a space of verbal attributes. The configuration of dyads based on VAME ratings mapped weakly into the composite timbral similarity circumplex, however. Another substantial experiment was carried out using adjectives derived from Piston's (1955) *Orchestration* (Kendall and Carterette 1993c). We hoped thereby to increase the ecological validity of the scales. In addition, a rating scale of 0 to 100 was used, rather than the negation of the attribute. The 3-D MDS solution accounted for 95.1% of the variance. From this and the principal components analysis (PCA), interpretations were: dimension 1, *power*; dimension 2, *strident*; dimension 3, *plangent*. For sake of legibility, only the 2-D MDS solution (accounting for 84.3% of variance) is shown (Fig. 9). Thus, by using a variety of techniques, e.g., PCA, cluster analysis, MDS, and attribute vector analysis, we obtained a superior and interpretable mapping of word to sound.

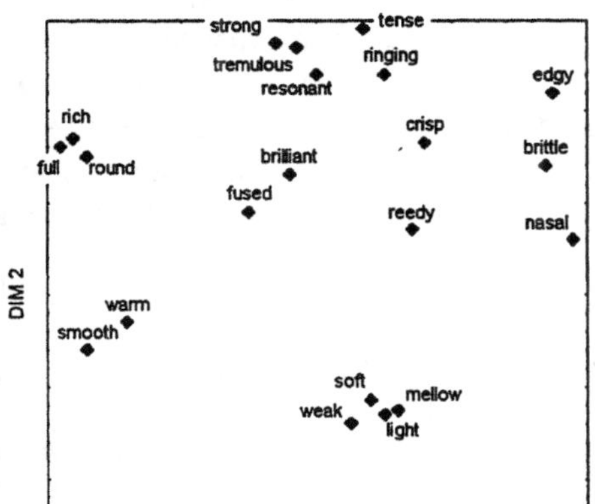

Fig 9. A plot of the first two dimensions (*D*) of a 3-D multidimensional scaling solution based on a 21 x 21 Piston-attribute correlation matrix. (From Kendall and Carterette 1993b)

Comparative Music Perception and Cognition

A final case shows our convergent approach in a very broad context (Carterette et al. 1993). In an effort to answer the question "Can certain metallophones, like bells and gamelan instruments, be assigned an unambiguous pitch?" we had to deal with many frames of reference and with a variety of experimental and mathematical methods. The study changed dynamically during its course as several partial answers converged on a larger view. We were led from an initial question about pitch perception to study the physics of metallophone bars. From the spectral properties of single bars we were led to study the musical relations among bars, thence to perceptual experiments, one on matching pitches, one on matching intervals. Little evidence was found that interval-sized patterns, even of small-cents deviation, were the major basis for perceived modes (pathets).[1]

There has been considerable controversy in the interpretation of data, variously obtained, regarding both the gamelan metallophone tuning and spectral and timbral characteristics (e.g., Hood 1954; Kunst 1973; Beurmann and Schneider 1989; Surjodiningrat et al. 1972). These acoustical and psychoacoustical facets of organological and musicological research have spawned additional controversies regarding the classification of Javanese *pathets* or modes. Many of the problems evaporate when the various frames of reference are appropriately linked.

We studied the comparative acoustical and psychoacoustical elements of the Javanese gamelan with a focus on tuning, vibrational characteristics and associated perceptual effects. In addition to relating acoustical, perceptual and musical frameworks to each other, we sought to clarify some common misconceptions about tuning modes, scales, and the assignment of pitch to metallophones which have inharmonic spectra. We found that the inharmonic modes of the spectra of the bars of the gamelan instruments *saron demung*, *saron barung*, and *saron peking* were described well by the standard physical theory of rectangular bars. Their makers have tuned the bars by shaping (filing, undercutting) so that harmonic partials also appear. The frequencies of the scales of the seven measured bars of each instrument (low 6, middle 1, 2, 3, 5, 6, high 1) were fit perfectly by an exponential function which implies a slightly stretched equipentatonic scale. Mean measured frequency intervals were ca. 246.5 cents with standard deviations of ca. 14 cents, relative to a perfect equi-pentatonic interval of 240 cents. Slightly stretched intervals with increasing frequency is universal to musicians and nonmusicians (Dowling and Harwood 1986).

---

[1] Successive intervals of the Western diatonic scale are unequal and for the major mode have the pattern *WWHWWWH*, where W = a whole-tone interval = $2H$, with H = a semitone interval ($H = 2^{1/12}$). Different (cyclical) patterns of *W*s and *H*s give different modes. Usually the pattern is written *LLSLLLS* (L = long, S = short) which masks the important fact that S is the modular interval of all modes. A typical Javanese seven-tone *pelog* scale has the complementary pattern *SSLSSSL*. From the *pelog* scale distinctly different pitch-class sets are built by choosing different subsets of five chroma (chroma = pitch class) which gives modes in the usual sense. But for *slendro* scales, there are only five tones per octave and usually all five tones are played, so that one "mode," or *pathet*, cannot be different from another because there is but a single interval size $P = 2^{1/5}$.

Results of perceptual experiments made in this study showed that, although many octave mismatches occur, both Indonesian and Western subjects generally matched a pitch chroma. In judging small-cents deviations from equipentatonic tuning, we found but one professional Indonesian musician of four who could reliably discern them.

The paradox of the existence of at least three different *pathets* (modes) in the face of equipentatonic tunings must be explained by factors outside of interval-size patterns like those of Western major and minor modes.[1] The basis of *pathets* lie in different probability distributions of chroma (pitch class) based on Markof chains of order 0, 1, 2 or higher and other imposed formulaic and hierarchical structures, which we call chromals. We believe that a number of other factors operate as well, for example, performance, orchestral composition, and intended, inherent timbral differences among gamelans. Indeed, the Javanese word *pathet* means "constrained." As put recently by Carterette et al. (1993),

> In Western music there is also an interaction of well-known melodic materials with the small-cents turning deviations of particular instruments, performers, and settings which create salient hallmarks. We suggest here that, in a fixed pitch orchestra of metallophones such as the gamelan, aspects of musical expression are embodied in the design of the instrument itself, exemplified in the pattern of tuning deviations. It seems to us that obvious functions and structures like tuning, which attract so much research as ends in themselves, lead us away from the deeper, multi-dimensional issues of musical communication (p. 395).

Conclusion

These results show how rating, identification, imaging, matching, similarity-scaling, and semantic-differential methods converge on the truth in the study of timbral combinations. It may also be necessary to adapt or recast old methods or to invent new methods.

We hope you are convinced of at least three main propositions:

1. What we learn in cognitive music research depends on what we ask and how we ask it.
2. Cognitive music research is driven implicitly and explicitly by assumptions, models, and theoretical constructs of both cognitive psychology and music.
3. Cognitive music research requires the use of multiple, convergent methods and the invention or derivation of new approaches.

References

Beurmann A, Schneider A (1989) Probleme und Aufgaben akustischtonometrischer Forschung in der Vergleichenden Musikwissenschaft. Acustica 69: 156-162

Bismarck G von (1974a) Timbre of steady sounds: a factorial investigation of its verbal attributes. Acustica 30:146-159

Bismarck G von (1974a) Sharpness as an attribute of the timbre of steady Sounds. Acustica 30:159-192

Butler D (1989) Mapping the listener's recognition of tonality: a delineation of the theory of intervallic rivalry. In: Proceedings of the 1st International Conference on Music Perception and Cognition. Japanese Society of Music Perception and Cognition, Kyoto pp 425-430

Butler D (1990) Response to Carol Krumhansl. Music Percep 7(3):325

Carterette EC, Kohl DV, Pitt MA (1986) Similarities among transformed melodies: the abstraction of invariants. Music Percept 3:393-409

Carterette EC, Kendall RA, DeVales S (1993) Comparative acoustical and psychoacoustical analyses of gamelan istrument tones. J Acoust Soc Jpn (E) 14(6):383-396

Dowling WJ, Harwood DL (1986) Music cognition. Academic, New York

Hood M (1954) The nuclear theme as a determinant of Pathet in Javanese music. J.B. Wolters, Groningen

Kendall RA (1986) The role of acoustic signal partitions in listener categorization of musical phrases. Music Percept 4(2):185-214

Kendall RA, Carterette EC (1990) Communication of musical expression. Music Percept 8:129-162

Kendall RA, Carterette EC (1991) Perceptual scaling of simultaneous wind instrument timbres, Music Percept 8:369-404

Kendall RA, Carterette EC (1992) Convergent methods in psychomusical research based on integrated, interactive computer control. Behav Res Methods Instrum Comput 24(2):116-131

Kendall RA, Carterette EC (1993a) Identification and blend of timbres as a basis for orchestration. Contemp Music Rev 9:51-67

Kendall RA, Carterette EC (1993b) Verbal attributes of simultaneous wind instrument timbres. I. Von Bismarck's adjectives. Music Percept 10(4):445-467

Kendall RA, Carterette EC (1993c) Verbal attributes of simultaneous wind instrument timbres. II. Adjectives induced from Piston's Orchestration. Music Percept 10(4):469-501

Krumhansl C (1990a) Tonal hierarchies and rare intervals in music cognition. Music Percept 7(3):309

Krumhansl C (1990b) Cognitive foundations of musical pitch. Oxford University Press, Oxford

Kunst J (1973) Music in Java: its history, its theory, and its technique, 3rd edn. Martinus Nijhoff, The Hague (Earlier English edition 1949; first published in Dutch as De Toonkunst von Java, 1934)

Surjodiningrat W, Sudarjana PJ, Susanto A (1972) Tone measurements of outstanding Javanese gamelans in Jogjakarta and Surakarta, (2nd edn. Gadjah Mada University Press, Jogjakarta

Fundamental Factors of Music Comprehension

Helga de la Motte-Haber

The dominant theory of perception and comprehension assumes that the mind works as a self-organizing system. Perceivers are active explorers of their environment and, hence, pick up information from it. But that information is not the only source of perception. Perception is linked to action which produces information. Environment and perceiving organism have a dialectical relationship which can be described by a paradoxical statement: Information processing is either the consequence of information or the source of that information. There is no direct mediating of environmental information. Inasmuch as perception implies the interpretation and completion of incoming stimuli, this is called comprehension.

One of the fundamental mechanisms of information processing concerns the spontaneous act of reduction. Information is reduced by gathering stimuli and focusing attention on them in such a way that they can be distinguised from a more diffuse environment or vague background. There are many examples for this well-known phenomenon of figure and ground which characterizes all perceptual experience, including vision, audition and so on. If there is a continuous tone of a constant pitch and loudness which is interrupted for a moment by a tone of different pitch or loudness we hear the brief tone as a figure on the ground of the continuous tone.

The perceptual properties of figure and ground are different. The figure tends to be better defined; the ground appears to be less well structured. Out of a chaos of physical stimuli we establish order. The mechanism of figure and ground creates a differentiation of the perceptual field. It allows a melody to be heard even though a whole orchestra is playing different parts. In this case additional principles of perceptual organization are in operation: a process of grouping parts to form wholes and a process of separating figure from ground by a contour. Being able to distinguish a figure at all usually depends upon the existence of a contour which separates it from a ground. This contour is clearly perceived as belonging to the figure, rather than to the ground, even though it is actually as a physical boundary common to both.

A figure may sometimes be recognized even when the physical contours do not completely define it. There are many examples of visual experiences in which a figure is clearly perceived despite the missing physical contours. However, the most prominent examples can be derived from music. Listening to the solo sonatas and partitas of J. S. Bach we form contours of melodies even though 32nd notes are being played. Similar to this example is a musical technique used by Luigi Nono that suggests a melodical flow which actually is based on an illusion of the listener (de la Motte-Haber 1985).

Forming a contour seems to be a basic principle of melodic perception. Several experiments of Dowling and Harwood (1986) show that the distortion of the intervals does not influence the identification of melody if the contour is preserved. However, the impression of a melody is changed if only the contour and not the size of the intervals is changed. It is impossible to hear a familiar folktune when the contour is destroyed by octave transpositions of some tones while others remain in their original position. Pitch and interval information are less important than contour information. The above-mentioned examples of Bach and Nono show that the listener also tries to create a contour.

Stimuli that are more similar to each other will have a greater tendency to be grouped. Similarity of stimuli means that they are alike in various physical attributes, such as intensity, color, size etc. For melodic perception this grouping mechanism is enhanced by similar frequencies. It is surprising that this mechanism is stronger than the basic mechanism of grouping by proximity (grouping by proximity is defined as a tendency to group stimuli that are in closer spatial or temporal proximity). Deutsch (1982) demonstrated in experiments with dichotic listening that pitches are grouped by similarity of frequency such that a scale is formed. It does not matter that the pitches are received and heard by different ears. This effect can also be demonstrated with stereophonic loudspeakers. Similarity seems to be a psychological, not a physical quality. The scale illusion demonstrates that a basic biological mechanism such as sound location has no relevance for musical information processing.

To what extent does listening depend on knowledge acquired by learning? Without knowledge the listener has a vague sense of music but no understanding of it, and without knowledge he is unable to comprehend the form of a musical piece. His spontaneous grouping mechanism do not help him to discover the form of a sonata. A musical illiterate is not disturbed when he listens first to the development section and recapitulation and the exposition is offered afterwards.

A recent experiment confirmed this effect. Beethoven's well-known sonata op. 49 no. 1 was played in the wrong order: first the development and recapitulation and then the exposition. The subjects for this study were 65 participants (age 20 - 63) of a conference concerned with the problem of musical form. They were asked whether they knew the piece or not and whether they would like to hear another version of it.

The results were surprising. Thirty subjects believed they knew the sonata but their knowledge was not deeply enough structured to discover the wrong order. Maybe the high number of 30 subjects is more an indication of an effect of social desirability. Ten of these subjects, however, make additional remarks expressing a feeling that something was wrong. The subjects (total 24) who did not know the piece were not able to succeed in this task. Only seven subjects who knew the piece very well noticed the wrong order of the sections. These data support the hypothesis that spontaneous grouping tendencies or other fundamental principles of information processing have less relevance for listening to music. Musical "Gestalt" is something hat has to be learned.

I wonder why so few rhythm studies have been conducted; since rhythm shapes the flow of musical information in more easily comprehensible units. Rhythmic information is probably more fundamental to music cognition than pitch information is. In tonal music, the metric weighting of the bars are the base of a hierarchical syntax defined by harmonic progression. Notably, Monahan (1984) has shown, in a multidimensional task, that rhythm is the major, first ordered dimension of musical perception. An experiment I conducted in the 1960s showed that, in contrast to a rhythmic mistake, a melodic mistake is more likely to be discovered in an adagio tempo than in an allegro. I think today that exact pitch-information processing probably progresses slowly, and a melodic mistake will be more readily noticed in a slower tempo. There is less effort involved in comprehending rhythms because the metric structure already divides music into meangingful units. Give a meter or a mode music can readily be grouped into a repeated-beat pattern which functions both as a perceptual (stimulated by physical cues) and conceptual (due to learned categories) entity. There are only a few musical examples in which no such grouping into equal patterns exists.

I have tested the hypothesis that the grouping of temporal qualities is fundamental for music cognition, regardless of pitch information, even though, of course pitch-grouping mechanisms can further enhance comprehension. An unknown Spanish folk tune and a monophoic section from the piece Structures by Boulez were selected for an experiment. In addition to the original versions six further versions were constructed by shifting pitches into another octave range and by transforming the rhythmic structure (Fig. 1). Each musical example was played twice. Thirty graduate students of musicology had to indicate whether the second version was the same or not the same, or whether they did not recognize it. Simple time-grouping enhances comprehension especially well (Table 1).

Table 1: Number of correct (+) and incorrect (-,0) answers

musical example	+	-	0
2	24	6	-
3	12	17	1
4	9	17	4
5	6	21	3
6	9	21	-
7	22	8	-
8	10	18	2
9	11	14	5

+, same; - , not the same; 0, do not know it

Our experimental results indicate that in a tonal example the destruction of both the melodic contour and the metric structure confuses the listener. But for a very complex series of pitches, simple time-grouping increases perception more than

Example 2

Example 3

Example 4

Example 5

Example 6

Example 7

Example 8

Example 9

Fig. 1. An unknown Spanish folk tune and monophonic section from *Structures* by Boulez. The *examples 2-9* were presented with this numbers (instead of names) in the experiment. The number 1 was not included because it would have been overly conspicuous

melodic contour. Thus, recognizing rhythmic patterns seems to be more important than recognizing melody. It facilitates the fundamental process of forming units and therefore easier perception.

There was, however, one really unexpected result with the Boulez example (example 6). which cannot be explained by complexity of metric structure or by pitch organization.The example has the contour and the rhythm of the folktune. At first, I was really disturbed that listeners could not recognize this version when it was played the second time. I repeated the experiments and got the same results. I carefully interviewed the subjects, who mainly mentioned the lack of any sense plus the impossibility to pick up this series of tones, which conflicted with their perceptual categories. Indeed, d-flat minor is suggested at the beginning both by a rhythmic and a melodic gesture. However, this impression seems to be totally wrong after the e and f sharp have sounded.

Considering all of our musical examples (but especially examples 2 and 6) I suppose that at a very low level of perception, a stylistic scheme is working. This scheme need not be a tonal one (even though it was a tonal one in example 6). It is a perceptual set (in the sense of the theory of Bruner and Postman) functioning as a hypothesis for the listener who has lost his orientation, since the incoming information conflicts with his previous information. In the case of example 6 such a conflict arose, but not in the nontonal series example. (It seems that such a conflict also arose in example 5.) Implicit acquired knowledge (a stylistic scheme) can be more important than structural organization - even sometimes inhibiting the recognition of "Gestalt" principles.

The assumption that a stylistic judgment is automatically a basic process of perception is also confirmed by everyday experience which shows that listeners use such stylistic schemes. When listening to the radio, one changes stations quickly if the style of music is incompatible with one's categorical system.

Conclusion

No doubt, every psychologist will agree that information processing can be the source of information. But there is a strong tendency in music psychology studies search for a natural musical grammar. The success of the book by Lerdahl and Jackendoff (1983) stimulated the research for fundamental grouping mechanisms. Recently Lerdahl (1988) dared to propose even cognition constraints on composition systems and he suggested rules for the composer. According to him a listener comprehends when the composer uses a grammar which is based on listeners' fundamental spontaneous grouping mechanisms.

This idea is as crazy as all other proposals of "universals" or "cross-cultural invariants" which I found in books on music psychology. The grammar of the listener depends on his knowledge. Contrary to Lerdahl's opinion, the listener comprehends if he uses the compositional rules which the composer invented. Basic biological mechanisms such as sound location have no relevance for musical coherence. Music is often compared with spoken language. Scientists are hoping to find identical or similar strategies of information processing in both fields. But the conditions for language comprehension are quite different than those for music comprehension. It would be, for example, impossible to violate basic biological mechanisms as such as sound location. The cocktail party effect demonstrates that understanding speech depends on the identification of sound location. The scale illusion mentioned earlier, however, as well as the phenomenon of extracting contour show that listening to music is based on learned categories. The human being is accustomed to listening according to the principle of frequency similarity, which is the basis rule for constructing a melody in tonal music. But this rule is not universally valid. There are famous examples in which this melodic rule is violated. Stockhausen withdrew the first version of the *Kontra-Punkte*. Comparison of the second version with the first (if by chance the score for the first version is available) reveals a punctual construction in the revised piece. Therefore the composer destroyed any reference to melodic contours. The work became more complicated and more interesting.

Musical structure is the result of aesthetic imagination. The listener has to learn this structure and as the experiment with the Beethoven sonata indicates: he has to learn it for every piece in a specific way.

References

de la Motte-Haber H (1985) Handbuch der Musikpsychologie. Laaber, p 101

Deutsch D (1982) The psychology of music. Academic, New York, pp 108,109

Dowling WJ, Harwood DL (1986) Music cognition. Academic, New York, pp133, 134

Lerdahl F (1988) Cognitive constraints on compositional systems. In: Sloboda J (ed) Generative processes in music. Clarendon, Oxford, pp 231-259

Lerdahl F, Jackendoff R (1983) A generative theory of tonal music. MIT Press, Cambridge, MA

Monahan CB (1984) Parallels between pitch and time. The determinants of musical space. Ph.D. dissertation, University of California Los Angeles

Beginning of Human Musicality

Hanuš Papoušek and Mechthild Papoušek

Introduction

In relation to the regulation of human behavior, natural selection has favored perception and processing of sounds and visible events to a degree at which their combinations allow not only a perfect ecological adjustment, but also a development of complex forms of communication and culture; thus, biological adaptation improved far beyond the limits observable in other mammals, primates included. Moreover, humans have become capable of creating a mental representation of their environment and course of life. In this internal world, humans can integrate life experience according to new, for instance, moral or hedonic, principles and can attribute new meanings to environmental events. Consequently, sounds coming from enviroment may be perceived and appraised - with a considerable individual variability - as pleasant, harmonious, or melodious, or disagreeable, wild, etc., although physical criteria for similar attributions may be problematic. Hence, musicality - whether in terms of sound qualities, or in terms of the listener's sensitivity to, knowledge of, and talent for music - may also be difficult to evaluate with physical measurements.

Yet, in spite of problematic definitions and assessments, musicality has proved significant not only to human cultural life or nonverbal mediation of human emotions, but also to human health and resistance against detrimental circumstances. For good reasons, therefore, musicality is becoming a matter of increased attention. Theoreticians wish to know more about its genetic and environmental determinants or its relation to brain functions. Clinicians study effects of music on immunological resistance or on the course of affective disorders. Parents would like to know more about early assessment of musicality for educational purposes, and teachers of music may ask whether musicality may be influenced by early educational interventions.

In problems of this sort, a closer look at the origins of musicality may help, particularly if not only historical, but also biological origins are considered. Biologists' interest concerns development of musicality during individual lifespans, during the evolution of entire species, and in comparisons with other species. Comparisons across animal species often reveal pointers to adaptive relevance in phenomena under investigation: precursors of human competences, qualitative universalities, or species-specific characteristics. To biologists, the question of adaptive significance is of crucial importance; they examine relations to essential life processes, such as communication, foreaging, reproduction, care for progeny, social organization, or territorial claims.

Music and Communication in Evolutionary Perspective

In this presentation, we wish to focus on interrelationship between musicality and communication which, from our psychobiological view, is important for understanding origins and adaptive significance of musicality. Communication as such is vital to living organisms: whatever affects means of communication may affect the survival probability in a given species. As a part of communication, musicality may have been tangled in an evolutionary selection process, producing significant biological consequences: for instance, selection of innate anatomical and behavioral predispositions for communicative competences; a universal presence of such predispositions across age, sex, or culture; early functioning of related competences during ontogeny; participation of nonconscious, intuitive forms of regulation, allowing fast and economic performance; coevolution of supportive behavioral programs in social counterparts. Needless to say, a better understanding of these aspects may help to interpret some of the aforementioned theoretical, clinical, and educational aspects of musicality. However, an excursion into the prehistorical past of music and musicality represents a step into a poorly documented realm of research where a lack of reliable evidence can often only be overbridged with logical deductions and constructs.

Generally, social communication is possible if two or more individuals can produce and perceive some signals, process them in nervous systems, and, eventually, attribute them some meaning. Signals may concern various perceptual modalities, such as tactile, vibratory, thermic, olfactory, auditory, or visual. The latter two dominate in communications among birds or primates. During evolution, the increased participation of vocal sounds in communication has been interrelated with the development of new structures in mammalian brains, and with new anatomical and functional changes in vocal tracts. Similar changes evolved in birds, too. Those changes have brought about predispositions for production of finely differentiated vocal sounds and their application for differential symbols of various events, needs, risks, or social interchanges. Vocal sounds mostly function in intimate interrelations with visually perceivable patterns of movement.

Nonverbal signals in communication in nonhuman animals or human infants have often been viewed as mere expressions of affective states. According to Bühler (1934), however, every such signal carries triple information: expression of affective state, representation of communicative context, and appeal addressed to social environment. The capacity of the nervous system determines to what degree each of the triple aspects is utilized for communicative purposes. In contrast to reptilian brains, additional limbic and hippocampic structures in mammalian brains link communication with more complex forms of care for progeny including emotional bonding, display of separation calls, and elementary forms of playful behaviors

(McLean 1990). The complexity of related behavioral expressions varies among mammalian species and reaches a particularly high level of complexity and plasticity in primates.

Differentiation in communicative vocal sounds concerns the same qualities that are cultivated in music: pitch, intensity, timbre, melody, rhythm, and harmony. Human speech includes additional phonetic qualities that allow producing consonants and syllables and chaining them into strings according to grammatical or syntactic rules. Thus, the vocal tract can serve as both an instrument for vocal communication and a musical instrument. Although human speech and vocal music may seem to represent two different communicative categories, transitions and interrelation between them are fluent. It is also difficult to say which of the two forms of communication appeared earlier during human evolution.

In nonhuman animals, birds in particular, rich variations of sound qualities exist in absence of words or other forms of abstract symbolization. While listening to them, humans may find them musical, although birds may perceive them as mere information about the partner's territorial claims or reproductive competence, as signals about available food, or as alarm signals of environmental dangers. In mammals, too, vocal communication may be melodic and yet its meaning may be as limited as in bird songs. For instance, gibbons belong to singing primates; however, their songs mediate in a relatively rigid manner information on circumstances such as mentioned in relation to bird songs (Marshall and Marshall 1976). There is no evidence that gibbons modify vocal sounds playfully or attribute musicality to them .

In humans, all the aforementioned predispositions have been selected during evolution in a particularly advantageous combination (Lieberman 1984). Brain functions have reached a level of complexity that is unparalleled in other animals: they allow complex cognitive operations with vocal symbols in verbal forms, and playful or creative modifications of acoustic elements in musical forms. Both forms can be shared with social environment either in direct performance, or stored in written symbols. The human vocal tract, particularly its supralaryngeal parts, offers outstanding options for fast and fine modulations of vocal sounds. A comparable coordination of gestural movements enables parallel expressions in nonvocal forms, for instance, in sign languages wherever necessary. Thus, humans can utilize vocal sounds in combination with bodily movements for an infinite repertoire of expressions reaching far beyond the limits of communication in other animals. This advantage gave rise to human cultures with unprecedented progress in accumulation and application of knowledge.

Comparative approaches have elucidated the origins of human communication in several aspects. First, they have shown that humans share some seemingly unique capacities with other animals. For instance, formation of abstract concepts has been demonstrated in rats (Tolman et al. 1946) and pigeons (Herrnstein and Loveland 1964), or the use of words in several monkey species (Seyfarth et al. 1980). Second, those human capacities with deeper biological roots depend not only on cultural determinants, but also on innate predispositions and nonconscious forms of neural regulation (Papoušek and Papoušek, 1987, 1991). Third, scientific attention has been drawn towards early developmental periods of human life with the assumption that their better knowledge might help in elucidating the process of evolution to which scientists otherwise have only a limited access. These three aspects are closely related to the development of musicality and, therefore, deserve special attention.

Musical Elements in Early Social Interaction

Human infants used to be viewed as altricial, due to slow development in locomotion. However, communicative capacities may have been favored as more adaptive during human evolution; in those capacities, human infants are now considered precocious, and their dependence on caregivers is viewed as advantageous for speech acquisition during dyadic interchanges (Papoušek and Papoušek, 1986).

The newborn enters the social environment with some perceptual acoustic experience from intrauterine life and can, for instance, identify the mother's voice (DeCasper and Fifer 1980). However, in the production of vocal sounds, the newborn's options are narrowly limited. The most prominent part of neonatal repertoire is the cry for which a sufficient innate program is available, including loud intensity and prolongated breathing. Conversely, for a quiet communication with the caregiver, the newborn can utilize only soft and short displays of fundamental voicing, the length and rhythm of which are primarily determined by the momentary rate of breathing (Papoušek and Papoušek 1981). The first observable step towards differential vocalization is a gradual prolongation of these sounds and increasing variations in pitch. In 6 to 8 weeks after the birth, the frequency of quiet vocalization increases, and they acquire a pleasant character of "cooing" (Papoušek 1992).

Around that age, the infant also starts producing distinct vowel-like sounds and the first consonant-like sounds (protoconsonantal margins, according to Oller and Eilers 1992), combines them in primitive articulations until fully vocalic sounds and improving articulation lead to marginal babbling. Between 5 and 10 months, the infant reaches the stage of canonical, well-formed, and reduplicated syllables, such as [dada] and [baba] which in some languages already represent real words. Infants can use them as meaningful words between 10 and 15 months. However, the use of melodic variations does not stop or decrease with the onset of speech-like elements. On the contrary, infants usually enrich variations in pitch, intensity, or other acoustic/musical features both during social interchanges with the caregiver, or on their own during monologues (Papoušek and Papoušek 1981). Observable signs of attentive concentration or of pleasure indicate intentional exploration or playful modification of new acoustic options. Vocal tract seems now to function not only as a communicative instrument, but also as the first naturalistic toy and/or musical instrument. Sooner or later, usually during the second year of life, the infant may start singing the first songs provided by social environment.

Studies on infant perceptual capacities also indicate innate predisposition for processing musical elements in vocal sounds. Fassbender (1992) demonstrated segregation of melodic entities from combinations of sounds and regrouping them in holistic ways in infants above six weeks of age. Other authors showed that 2-to-3-month olds can perceive and process melodic segmentation of speech (Karzon and Nicholas 1989; Trehub 1990) and produce rhythmic or melodic patterns in vocalization (Kelly and Sutton-Smith 1987; Lynch et al. 1992). Preference for intonations used in infant-directed speech was found in 1-month old babies (Cooper and Aslin 1989), ability to imitate pitch and pitch contours in 2-month old babies (Papoušek and Papoušek 1989; Uzgiris 1972). In many more studies, perception and processing of harmonic or rhythmic structures was demonstrated in infants above 3 or 6 months of age at which musical stimulation occupied an increasing proportion of parental repertoire (for survey see Trehub 1990).

Thus, during ontogeny, infants first become capable of modifying melodic elements in vocal sounds for communicative purposes and only then they start using phonetic articulation, syllables, and words. If we take analyses of preverbal vocalization as a peephole into phylogeny, the observed sequence would indicate that a communication based upon melodic contours developed prior to verbal communication. Melodic qualities of vocal sounds are obviously easier to detect in noisy environments or in communication across larger distances. Therefore, populations living under unfavorable acoustic circumstances still utilize melodic features, for instance in yodelling or in whistle languages. Consequently, two alternative hypotheses should be considered. Either some combination of melodic features in voice with facial and manual gestures preceded verbal communication during evolution, or both melodic and verbal forms of communication evolved simultaneously and were utilized differentially, according to environmental acoustic circumstances.

Supportive Guidance to Early Musicality

In biological views, involvement of a behavioral tendency in some relevant adaptive process leads to selection of specific innate predispositions. Innateness is assumable if predispositions are rather universal across age, sex, or cultures, emerge early during ontogeny, depend on nonconscious forms of behavioral regulation, and if their evolution has been paralleled by a coevolution of supportive behavioral tendencies in social environments. Similar characteristics might be expected in relation to both verbal and musical forms of communication if their adaptive significance is valid. These consequences would be relevant to the question of how far the development of early musicality may depend on maturation or learning.

From our own former studies, we know that infants can learn from the first days of life, particularly in situations where they can control some contingent environmental changes. The earliest learning processes are slow, but accelerate rather rapidly during the first three months, due to both maturation and experience in learning situations (Papoušek, 1977; Papoušek and Papoušek 1984). To arrange such learning situations, experimenters have to follow certain "didactic" principles: they have to use simple forms of stimulation and contingency, choose infants' optimal state of wakefulness, and motivate infants with effective reinforcements.

Surprisingly, microanalyses of maternal behaviors have shown that mothers provide best-fitting learning situations for support to infant progress in vocal development (Papoušek and Papoušek 1987). In infant-directed speech ("motherese"), mothers reduce the structure of speech prosody into slower, simpler, exaggerated and repetitive patterns (Fernald 1992; Papoušek, et al. 1985), adjust the intensity of stimulation to infants' behavioral states, and reward every attempt in infants to join a dialogue with pleasurable affective reinforcements. Strikingly prolongated and melodically displayed vowels are prominent and universal features of infant-directed speech in various cultures (Papoušek and Papoušek, 1991; Papoušek et al. 1991). One can hardly imagine better models for guiding the infant towards a dialogic exploitation of the first available communicative opportunities - the use of melodic elements. However, we also found and verified that mothers were unaware of their effective teaching activities (Papoušek and Papoušek 1987). Their efficiency was documented in a prospective comparison of twin and singleton infants by Bornstein (1985). Twins received less didactic interventions from mothers and were significantly slower in development of speech than singletons.

Conclusions

The fact that a part of parental care is based upon nonconscious, intuitive behavioral tendencies indicates a participation of innate behavioral programs, such as is common in relevant adaptive behaviors in mammals like, for example, the care for their young. However, in other mammals, primates included, parental support usually concerns other than communicative behaviors; typically, parents may support the development of locomotion. Humans seem to be unique in supporting that behavioral repertoire which represents their most effective means of biological and cultural adaptation - verbal communication.

The present advances in research on infant perception and cognition make it evident that this coevolution also concerns predispositions for infant musicality. In other words, infants seem to be biologically predisposed for musicality and the development of their elementary musical competence seems to be nourished by intuitive parental caregiving. Interestingly, in this case of support nobody thinks of testing infant talent for musicality; nobody argues about whether to support musicality so early after the birth: all caregivers unintentionally give this support under normal conditions, and thus all infants obtain elementary "musical education", however, only up to the beginning of verbal age. Then, cultural determinants increasingly take the lead and can override biological determinants to a degree which means either an end of musical interaction or a beginning of an overexaggerated flood of noninteractive musical stimulation that finally leads to selfprotective aversion in the infant, if not to a hearing deficiency.

References

Bornstein MH (1985) How infant and mother jointly contribute to developing cognitive competence in the child. Proc Nat Acad Sci U S A 82:7470-7473

Bühler K (1934) Sprachtheorie. Fischer, Jena

Cooper RP, Aslin RN (1989) The language environment of the young infant: implications for early perceptual development. Can J Psychol 43:247-265

DeCasper AJ, Fifer WP (1980) Of human bonding: newborns prefer their mothers' voices. Science 208:1174-1176

Fassbender C (1992) Auditory grouping and segregation processes in infancy. Kaste, Norderstedt

Fernald A (1992) Meaningful melodies in mothers' speech to infants. In: Papoušek H, Jürgens U, Papoušek M (eds) Nonverbal vocal communication: comparative and developmental approaches. Cambridge University Press, New York, pp 262-282

Herrnstein RJ, Loveland DH (1964) Complex visual concept in the pigeon. Science 146:549-551

Karzon RG, Nicholas JG (1989) Syllabic pitch perception in 2to 3-month-old infants. Percept Psychophys 45:10-14

Kelly L, Sutton Smith B (1987) A study of infant musical productivity. In: Peery JC, Peery IW, Draper TW (eds) Music and child development. Springer, Berlin Heidelberg New York, pp 35-53

Lieberman P (1984) The biology and evolution of language. Harvard University Press, Cambridge MA

Lynch MP, Oller DK, Steffens ML (1992) Phrase-like rhythmic units in prelinguistic vocalization. 8th International Conference on Infant Studies, May 7-10, Miami

Marshall JT Jr, Marshall ER (1976) Gibbons and their territorial songs. Science 199:235-237

McLean PD (1990) The triune brain in evolution: role in paleocerebral functions. Plenum, New York

Oller DK, Eilers RE (1992) Development of vocal signaling in human infants: toward a methodology for cross-species vocalization comparisons. In: Papoušek H, Jürgens U, Papoušek M (ed) Nonverbal vocal communication: comparative and developmental approaches. Cambridge University Press, New York, pp 174-191

Papoušek H (1977) Entwicklung der Lernfähigkeit im Säuglingsalter. In: Nissen G (ed) Intelligenz, Lernen und Lernstörungen. Springer, Berlin Heidelberg New York, pp 75-93

Papoušek H, Papoušek M (1984) Learning and cognition in the everyday life of human infants. Study Behav 14:127-163

Papoušek H, Papoušek M (1986) Structure and dynamics of human communication at the beginning of life. Eur Arch Pychiatr Neurol Sci 236:21-25

Papoušek H, Papoušek M (1987) Intuitive parenting: a dialectic counterpart to the infant's integrative competence. In: Osofsky J (ed) Handbook of infant development, 2nd edn. Wiley, New York, pp 669-720

Papoušek H, Papoušek M (1991) Innate and cultural guidance of infants' integrative competencies: China and the United States. In: Bornstein MH (ed) Cultural approaches to parenting. Erlbaum, Hillsdale, pp 23-44

Papoušek M (1992) Early ontogeny of vocal communication in parent-infant interactions. In: Papoušek H, Jürgens U, Papoušek M (eds.) Nonverbal vocal communication: comparative and developmental approaches. Cambridge University Press, New York, pp 262-282

Papoušek M, Papoušek H (1981) Musical elements in the infant's vocalizations: their significance for communication, cognition, and creativity. In: Lipsitt LP (ed) Advances in infancy research, vol 1. Ablex, Norwood, pp 163-224

Papoušek M, Papoušek H (1989) Forms and functions of vocal matching in interactions between mothers and their precanonical infants. First Lang 9:137-158

Papoušek M, Papoušek H, Bornstein MH (1985) The naturalistic vocal environment of young infants: on the significance of homogeneity and variability in parental speech. In: Field T, Fox N (eds) Social perception in infants. Ablex, Norwood , pp 269-297

Papoušek M, Papoušek H, Symmes D (1991) The meanings of melodies in motherese in tone and stress languages. Infant Behav Dev 14:414-440

Seyfarth RM, Cheney DL, Marler P (1980) Vervet monkeys alarm calls: semantic communication in a free-ranging primate. Anim Behav 28:1070-1094

Tolman EC, Ritchie BF, Kalish D (1946) Studies in spatial learning. II. Place learning versus response learning. J Exp Psychol 36:221-229

Trehub SE (1990) The perception of musical patterns by human infants: the provision of similar patterns by their parents. In: Berkley MA, Stebbins WC (eds) Basic mechanisms. Wiley, New York, pp 429-459 (Comparative perception, vol 1)

Uzgiris IC (1972) Patterns of vocal and gestural imitation in infants. In: Moenks FJ, Hartup WJ, deWitt J (eds) Determinants of behavioural development. Academic, New York, pp 467-471

Expressive Intention and Performance

ALF GABRIELSSON

Theoretical Background

This paper discusses how intentions concerning the emotional character of a piece of music affect the performance of the piece. In other words, what does a performer do to generate the intended emotional character of the music? We will first present two theories on music and emotion and then turn to some empirical investigations.

Susanne Langer

One of the most influential theories concerning music and emotion was proposed by the philosopher Susanne Langer (1953, 1957). A basic idea in her theory is that there is an isomorhpism between the structure of music and the structure of feelings, "music is a tonal analogue of emotive life" (Langer 1953, p. 27), "music is...formulation and representation of emotions, moods, mental tensions and resolutions" (1957, p. 222). The latter statement was a reformulation of a well-known statement by Richard Wagner concerning the expressiveness of music. Langer invoked many other authors as well. Thus Wolfgang Köhler (1947, cited in Langer 1957, p. 226) found musical terms useful for describing the forms of mental life: "Quite generally the inner processes, whether emotional or intellectual, show types of development which may be given names, usually applied to musical events, such as: crescendo and diminuendo, accelerando and ritardando". Carroll Pratt (1931, cited in Langer, 1957 p. 244) remarked concerning the auditory character of music that "[they] are not emotions at all. They merely sound the way moods feel..." This statement was later succinctly put "Music sounds the way moods feel" (Winner 1982, p. 211) Furthermore, in discussing the differences between language and music, Langer stated that "because the forms of human feeling are much more congruent with musical forms than with the forms of language, music can reveal the nature of feelings with a detail and truth that language cannot approach." (Langer 1957, p. 235).

The formal properties that may be similar in music and in human feeling are, for example, "forms of growth and attenuation, flowing and stowing, conflict and resolution, speed, arrest, terrific excitement, calm, or subtle activation and dreamy lapses" (Langer 1953, p. 27), "patterns of motion and rest, of tension and release, of agreement and disagreement, preparation, fulfilment, excitation, sudden change, etc." (Langer 1957, p. 228). These properties are intuitively appealing to many of us and often appear in various writings on music. However, Langer herself did not go into much musical detail in order to explicate these factors. As a kind of hint she approvingly referred to the study by Huber (1923) on the expressiveness of short pitch patterns ("*musikalische Elementarmotive*"). Huber presented his listeners with pitch patterns of two or three tones, with uniform timbre, volume, and duration, and asked them to describe what the motive "told them" ("*Was sagt mir das Motiv?*") with regard to qualities, emotional characters, associations, meanings, somatic effects, and other things. The results showed convincingly "how many factors of possible expressive virtue are involved in even the simplest musical structure" (Langer 1957, p. 231), such as tone color, width of tonal intervals, direction of tonal movement, subjective accents, impressions of relatedness among tones, and others. The implication is that in real music the number of factors is larger still and their relations even more complex.

Manfred Clynes

Another starting point may be taken in Manfred Clynes's "sentics" theory (Clynes 1977; Clynes and Nettheim 1982). Clynes emphasized that emotion and its expression form a unit, an integrated system. However, an emotion ("sentic state") may be expressed by different output modalities, such as "gestures, tone of voice, facial expression, a dance step, musical phrase, etc." (Clynes 1977, p 18). To get an easily recorded measure of expression, Clynes constructed a so-called sentograph, a device on which subjects express different emotions by pressing a slightly elastic button with the middle finger. The pressure patterns ("sentograms") are recorded, and Clynes claimed that there are distinctly different patterns for different emotions, such as joy, grief, anger, hate, and others.

The same device can be used to study listeners' experiences of pulse and rhythm in music. While listening to the music the subject presses the sentograph button in accordance with how he feels the pulse or motion character of the music. Examples of such "music sentograms" appear in Clynes and Walker (1982) and in Gabrielsson (1993). The shape of these sentograms can then be compared with sentograms for different emotions, and thus provide a nonverbal method of exploring the emotional expression of the music.

(Of course, the description of Langer's and Clynes's theories here is incomplete and selective. For a complete discussion, refer to the original texts.)

Emotion, Motion, and Music

Despite all differences between Langer's and Clynes's approaches, their theories are similar in pointing to an intimate relation between the structure of emotions and the structure of music. Langer does this at a rather "global" level, referring to common properties as forms of growth and attenuation, motion and rest, tension and release, conflict and resolution, preparation, fulfilment, etc., whereas Clynes focuses on certain "basic emotions" (joy, grief, anger, hate, reverence, love, and sex) and their possible expression in pitch and rhythm patterns.

We may also generally notice the fact that we behave/move differently in different moods, as reflected in expressions like "to jump for joy", "sink in despair", "tremble for fear" etc. That we likewise move differently when listening to music of different emotional characters can also be easily observed. Exciting music may evoke vigorous movements, whereas other music may make us dance, rock, or swing. Still other music may bring us to motionless stillness. The intimate relationship between music and motion, overt as well as experienced, has been pointed out, in one way or another, by a multitude of authors ever since ancient Greece. For example, Truslit (1938) claimed that to shape music is essentially to shape motion, Langer (1953) wrote that "the elements of music are moving forms of sound" (p. 109, in referring to Hanslick's famous description of music as "*tönend bewegte Formen*"), and the composer Roger Sessions (cited in Clynes and Walker 1982, p. 172) stated that "we experience music as movement and gesture". The present author has repeatedly emphasized the motional - emotional aspects of music experience and performance (Gabrielsson et al. 1983; Gabrielsson 1986, 1988).

In summary, we may consider emotion, motion, and music as being isomorphic.

Empirical Investigations

Earlier Work

Empirical work in direct connection to the theories discussed above is still meager. However, there are numerous other investigations in which listeners were asked to describe the emotional character of different pieces of music using checklists of adjectives, semantic differential techniques, etc. In many of these studies an attempt was also made to relate the described emotional characters to various elements of the musical structure. Reviews of such investigations may be found in Dowling and Harwood (1986), Jörgensen (1988), and Radocy and Boyle (1988). Some fairly typical results are as follows (Rigg 1964):

- Serious and solemn music is slow, low pitched, and avoids irregular rhythms and dissonant harmonies.
- Sad music is likewise slow and low pitched and further apt to be in minor mode and to contain dissonance.
- Calm music is slow and soft and apt to contain little dissonance.
- Happy music is fast, high pitched and in major mode and contains little dissonance.

- Exciting music is fast and loud and apt to contain dissonance.

It should be noted that the music examples used were taken almost exclusively from Western art music or popular music. Furthermore, the description of the emotional character as being "happy," "sad," "calm" etc. only represents a kind of "average" or dominating emotional expression during the piece. However, the emotional character in music often changes within the piece - more or less gradually or suddenly - and so more process-oriented research techniques would be desirable. An interesting example is the work by Nielsen (1983) in which he studied experienced tension in two pieces of music. The listeners used a pair of tongs with a spring resistance to indicate the experienced tension – the more tension, the harder they pressed the tongs and vice versa. The results were displayed as "tension curves" along the time axis (bars) of the music and could be related to a multitude of structural elements in the music, such as loudness level, melodic contour and rhythmic and harmonic complexity. Although Nielsen did not refer to Langer, this study comes close to studying the types of more global and time-varying properties suggested by Langer, in this case tension and release.

Factors in Performance

There are thus many factors in the musical structure that can contribute to the generation of a certain emotional character, such as loudness (dynamics), tempo, pitch level, rhythm pattern, melodic pattern, harmonic progression, tonality, and so forth. In the case of composed music these factors are indicated in the score. It is well - known, however, that a conventional score is incomplete in many respects, leaving much to the performer's interpretation. A host of studies on the microstructure of musical performance, dating from the very beginning of this century, demonstrate a multitude of "deviations" - in timing, dynamics, and intonation - from the nominal values in the score. These studies are reviewed in Gabrielsson (in press). It is generally considered that such deviations are essential for attaining certain desired qualities of musical performances. For instance, Seashore (1938, p. 29) claimed that "the medium of musical art lies primarily in artistic deviation from the fixed and regular."

Usually the deviations in timing, dynamics and intonation are considered means of clarifying the structure of the piece to the listener, for instance, by executing ritardandos at the end of phrases, inserting micropauses between phrases or other sections, provide accents for making the meter obvious etc. There is sometimes also a general reference to the "expressive purposes" of such deviations, however without any further specifications of what that would mean. I am convinced that some deviations are made for structural purposes but others for the sake of the proper motional - emotional character (and still others maybe for both purposes). Evidence for this was suggested in Gabrielsson et al. (1983), for instance, by showing how patterns that were identical in the notation were nevertheless performed differently in pieces of different emotional character.

This is no news whatsoever. Every musician of at least some standard knows, more or less consciously, that he may vary the tempo, articulate tones differently, emphasize some tones, play with vibrato or not, make the tone onset abrupt or gradual, sharpen or flatten the tone pitch, etc. in order to affect the emotional character of the piece. This is simply common musical knowledge.

Furthermore, a musician may play the same piece differently on different occasions - change his interpretation of the piece - and different musicians may differ much among themselves in the performance of the same piece. A large part of this intra- and interindividual variation is due to differences in understanding and internal representation of the music, including its proper emotional character. Thus, even if the same score is used, there will be differences in performance. Any system that attempts to formulate rules about musical performance based exclusively on the score is doomed to failure. The performer's representation of the piece and the resulting intentions are decisive. The (micro) structure of musical performance can never be fully understood without taking the performer's intentions, and the listener's experience, into account.

There is, however, little empirical research on this matter. Nakamura (1987) and Palmer (1989) studied the effects of performers' intentions concerning structural charcteristics, such as phrasing, tempo, and dynamics. Kendall and Carterette (1990) asked for performances "without expression," "with appropriate expression" and "with exaggerated expression". Askenfelt (1986) asked violinists to play a tender and an aggressive version of a Beethoven theme. The measurements of the performances revealed differences due to the prescribed characters; for instance, the aggressive version was louder and the changes in loudness were more abrupt than was the case in the tender version.

New Experiments

Methods

In order to investigate the effects of intentions concerning the emotional character on the microstructure in musical performance we are presently conducting a series of studies in which musicians are asked to perform short pieces of music. Each piece should be played in different versions, namely, so that it sounds happy, sad, solemn, angry, soft/tender, and indifferent (without expression).

So far the experiments have comprised performances on the flute, violin, electric guitar, and synthesizer and a vocal performance. The performers were, in general, professional musicians. Only monophonic performance (the melody) was used in order to facilitate recording and measurements. The performances took place in a laboratory with reasonably good acoustics (all recording and measuring equipment was in another room) or in a recording studio (for the flute, the violin, and the singer). The recordings of the synthesizer performances were made directly to computer memory via MIDI. All other performances were recorded on tape, electrically for the electric guitar and via microphone in the remaining cases. Video recordings were made as well. The tape - recordings were then stored in computer memory using one or two sampling systems in our laboratory. Sampling frequency up to 50 kHz can be used. This gives excellent possibilities to accurately analyze timing and dynamics as well as intonation.

The pieces were usually short. Most of them were well known but some were unknown, and one piece was composed for this investigation. When pieces were well - known, the performers were encouraged to play them by heart. For the unknown pieces the musicians were provided with music, but it could sometimes be done without after some practice. It was emphasized that any means whatsoever were allowed in order to give the desired emotional character, except that the pitches of the melody must be retained. The performers were thus free to do what they wanted as regards tempo, rhythm, dynamics, articulation, phrasing, etc.

After explanation of the task the musicians were given opportunity to practice on the respective instrument (their own instrument except for those who played the synthesizer). They had no difficulties in understanding or performing the task and generally found it interesting and thought-provoking. Each version was to be played twice with about a 2 s break in between. They could repeat a performance as many times as they wanted until they were satisfied (in most cases one or two attempts were enough). There was a relaxed atmosphere throughout. Afterwards the performers were interviewed regarding what means they had used in order to bring about the desired emotional characters, which character was the easiest and the most difficult to execute, and generally to tell about the experiences and ideas occurring to them during the experiment.

In one experiment the subjects were also asked to perform the different versions on the sentograph without any sound. They thus pressed the sentograph button imagining playing the respective version.

Results

Only a glimpse of the results can be given here. There were obvious differences among the different performance versions (happy, sad, solemn, angry, soft/tender, without expression) of each melody. There were also many differences among the performers. For instance, some of them tried to vary the expression while mainly remaining within the limits of the notation, whereas others took advantage of the freedom they had to change anything they wanted except the pitches. Thus some performers radically changed the structure of the music by using another meter, changing rhythmic patterns, and varying the tempo in rather extreme ways. This happened especially for the angry versions. In order to see which versions actually conveyed the intended emotional character to listeners we conducted several listening tests.

The results with regard to tempo and loudness were mainly as could be expected. Angry and happy versions were played faster than sad, soft, and solemn versions.

a

b

Fig. 1. **a** Notation of "Oh, my darling Clementine."
b The beginning of a melody composed for this study.

With regard to loudness level, angry versions were played loudest and soft versions softest (of course).

With regard to the microstructure in the performances some interesting results were as follows. In the tune "Oh, my darling Clementine" there is a recurring rhythm pattern, that is ♪. 𝅘𝅥𝅯 ♩ ♩ (see Fig. 1). The dotted pattern ♪. 𝅘𝅥𝅯 appears in upbeat position. The nominal ratio between the dotted eighth note and the sixteenth note is 3:1. As known from earlier performance studies (Gabrielsson et al. 1983), there may be considerable deviations from this ratio. The dotting may be sharpened, that is, have a ratio larger than 3:1, or softened with a ratio smaller than 3:1. In the present performances the number of dottings close to the nominal ratio (from 2.8:1 to 3.2:1), appeared in about only 12 % of the cases. The dottings were generally softened in the soft versions with ratios varying from about 1.75:1 to 2.5:1. In fact, two performers found it necessary to completely abandon the dotting and instead perform the pattern as two equally long (eighth) notes. On the other hand, dottings in the solemn and angry versions were often sharpened to ratios well above 3:1. In the happy version dottings were moderately softened with ratios varying from about 2.4:1 to 2.85:1.

Another interesting difference between the versions appeared with regard to the articulation of the tones. As seen in Fig. 2, the performance was mainly legato in the soft and solemn versions, sometimes even legatissimo (that is, the key on the synthesizer was not released until some time after the depression of the key for the next tone). On the other hand, the happy and angry versions show a variation between staccato and legato performance, which should be studied in relation to the structure of the tune.

A similar phenomenon appeared with regard to the loudness level. The angry and happy versions had the largest dynamic range, while the soft version was soft throughout with little dynamic range.

How the onset of a tone can be shaped, depending on the intended emotional character, is seen in Fig. 3. It shows the amplitude envelopes for the first few tones in the melody in Fig. 1b performed by a violinist. It is instructive to look at the onset of the very first tone in each version. In the angry version there is an extremely rapid onset, the amplitude reaches its peak value practically immediately - and in listening one can really hear the "bite" in this attack. By contrast the sad and soft versions show a slow, gradually rising onset. The onset in the happy and solemn versions are in between these extremes.

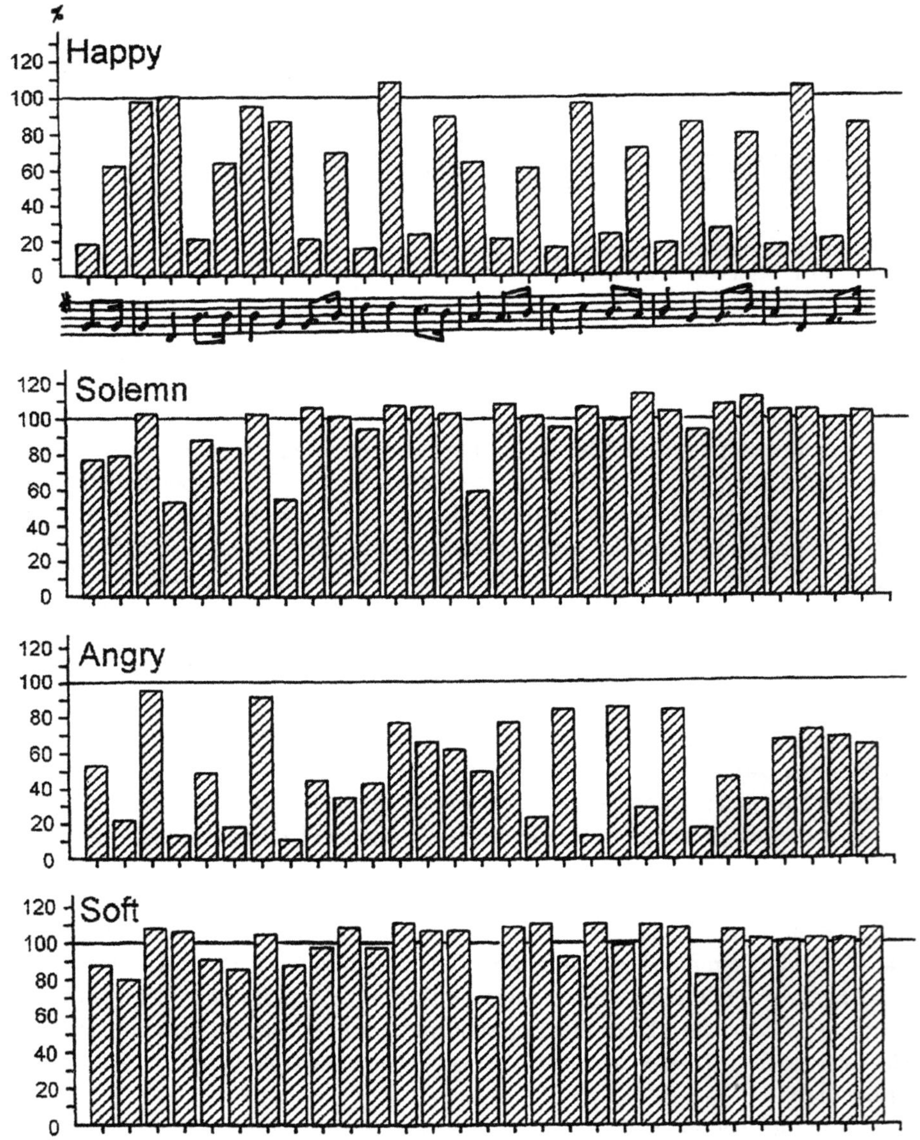

Fig. 2. Articulation of each tone in different versions of "Oh, my darling Cementine" performed on a synthesizer. The *height of the bar* above each note reflects the degree of legato performance. The *horizontal line* at 100 designates complegato performance. *Bars* higher than this represent legatissimo, and *short bars* represent staccato

The performance of the versions without expression showed many interesting features. Only two are mentioned here. In performing the "Clementine" tune, one performer simply made all tones equally long (say, as eighth notes) and equally loud, only retaining the pitches as he had to do - and it certainly sounded

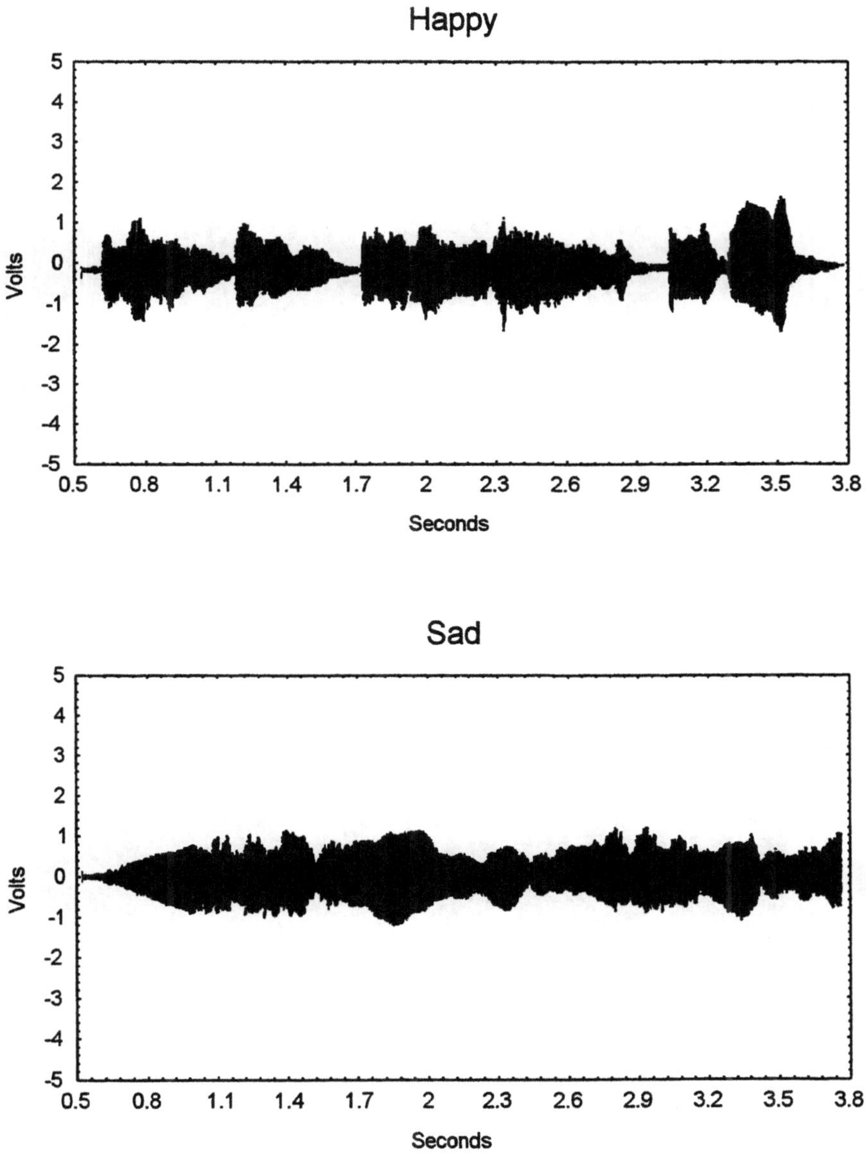

Fig. 3. a, b

expressionless. And when the violinist were to play the tune in Fig. 1b without expression, he deliberately skipped the vibrato that he used in all other versions, thus attaining a very different, "cold" timbre.

A final example of a somewhat different character is given in Fig. 4. It shows three pressure patterns ("sentograms") of the beginning of "Oh, my darling Cle

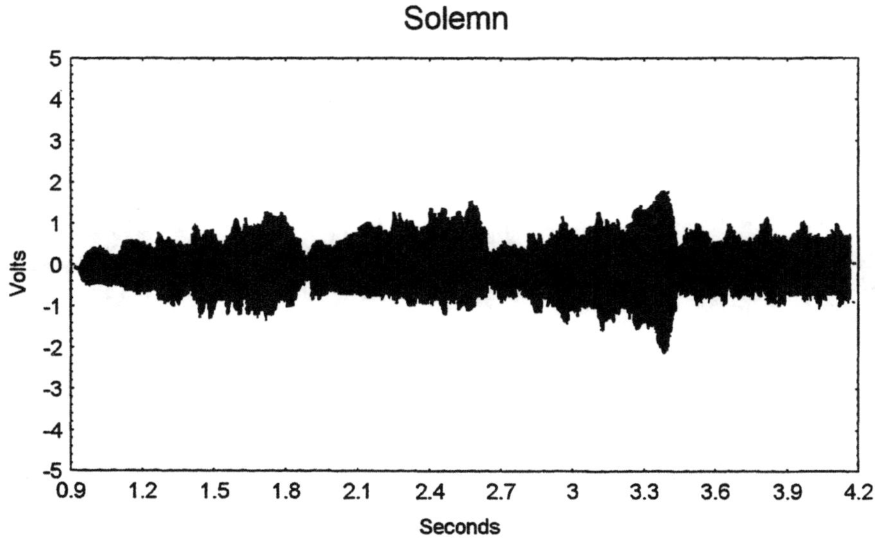

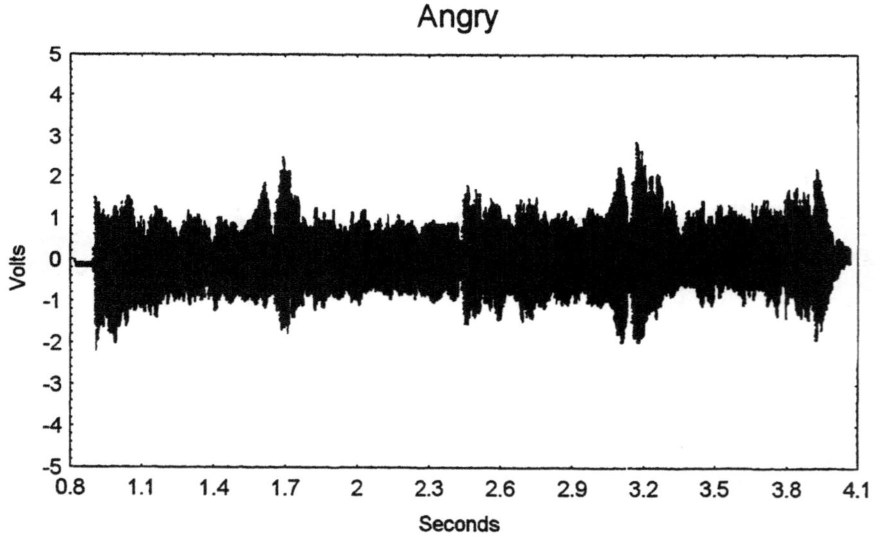

Fig. 3. c, d

mentine," that is, recordings of the vertical pressure exerted on the sentograph button when "playing" the tune on the sentograph. Three excerpts are shown: the happy, angry, and soft versions by the same performer. The angry version displays the largest amplitude of the pressure and distinct transitions between the "tones" (and rests). The (approximate) notation below shows how the performer changed

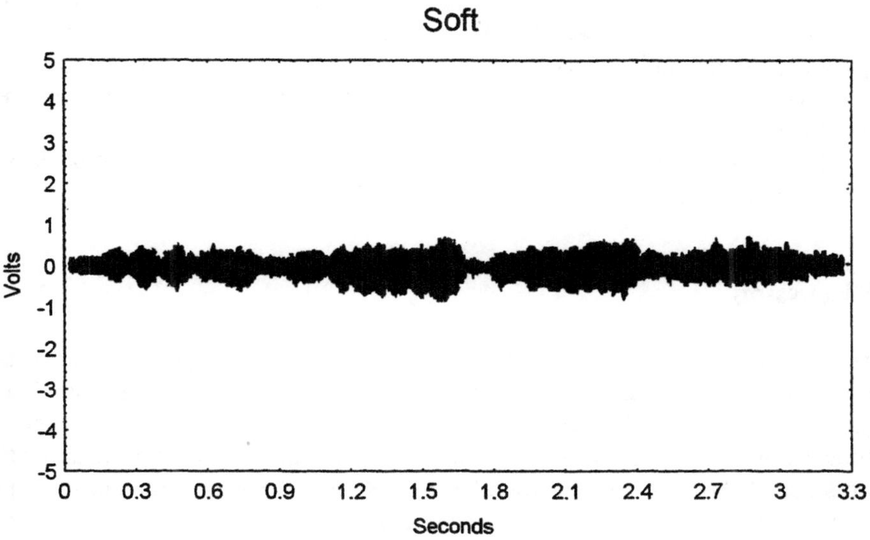

Fig. 3. a - e Amplitude envelope for the first tones in different versions of the melody in Fig. 1. b. The happy version displays the envelope for the first six tones, the other versions for the first three or four tones. There are obvious differences in the shaping of the tone envelopes among the different versions

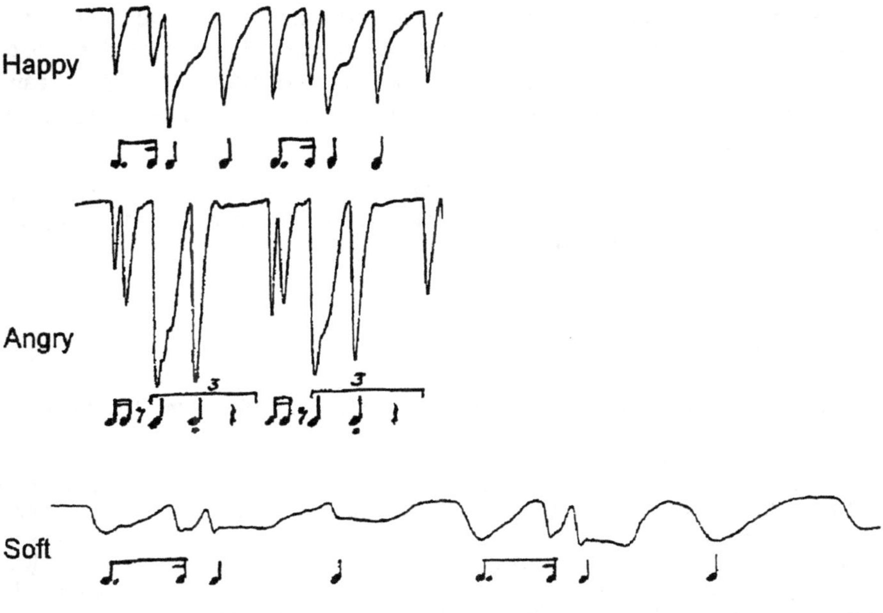

Fig. 4. Vertical pressure patterns in sentograph performance of the beginning of "Oh, my darling Clementine" in happy, angry and soft versions.

the rhythmic structure of the tune in this version. The happy version has about the same tempo as the angry one and likewise distinct transitions between "tones" but generally lower amplitude. The soft version shows the slowest tempo, the smallest amplitude, and a more legato character with gradual, "rounded" transitions betwen the "tones."

The complete results from our ongoing series of experiments will be published elsewhere. Some results were presented in Lindström (1992).

Discussion

This chapter has focused on a theory concerning an isomorphism between the structure of emotions and the structure of music, as proposed by Langer and Clynes (and suggesting that the isomorphism should include motion as well). There is considerable empirical evidence in support of this theory, although most of the works cited above do not directly refer to it. Here we have attempted to study the isomorphism with special regard to the microstructure in music performance, that is, how musicians use factors such as deviations in timing, articulation, dynamic range, amplitude envelopes (tone onsets), vibrato, and other things to generate various emotional characters. Although Langer in her examples of formal properties common for emotions and music mainly mentioned rather "global" factors, it seems that the isomorphism principle is applicable to all levels, from the shaping of the individual tone up to large musical forms. (A hint in this direction may be seen in Langer's discussion of Huber's (1923) investigation cited earlier.) Much empirical work remains in order to investigate this. While part of it can be done using conventional experimental methods, we should not hesitate to use other means, e.g., various phenomenological methods when necessary, especially regarding entities at the more global levels.

Acknowledgments. The author wants to express his gratitude to Erik Lindström, Guy Madison, Anna-Karin Gullberg, and Patrik Juslin for valuable cooperation. The research project was supported by The Bank of Sweden Tercentenary Foundation.

References

Askenfelt A (1986) Measurement of bow motion and bow force in violin playing. J Acoust Soc Am 80:1007-1015

Clynes M (1977) Sentics. The touch of emotions. Anchor Doubleday, New York

Clynes M, Nettheim N (1982) The living quality of music: neurobiologic patterns of communicating feeling. In: Clynes M (ed) Music, mind and brain. The neuropsychology of music. Plenum, New York, pp 47-82

Clynes M, Walker J (1982) Neurobiologic functions of rhythm, time, and pulse in music. In: Clynes M (ed) Music, mind and brain. The neuropsychology of music. Plenum, New York, pp 171-216

Dowling WJ, Harwood DL (1986). Music cognition. Academic, New York

Gabrielsson A (1986) Rhythm in music. In: Evans JR, Clynes M (eds) Rhythm in psychological, linguistic and musical processes. Thomas, Springfield, pp 131-167

Gabrielsson A (1988) Timing in music performance and its relations to music experience. In:

Sloboda JA (ed) Generative processes in music. The psychology of performance, improvisation, and composition. Clarendon, Oxford, pp 27-51
Gabrielsson A (1993) The complexities of rhythm. In: Tighe TJ, Dowling WJ (eds) Psychology and music. The understanding of melody and rhythm. Erlbaum, Hillsdale, pp 93-120
Gabrielsson A (in press) Music performance. In: Deutsch D (ed) The psychology of music, 2nd edn. Academic, New York (in press)
Gabrielsson A, Bengtsson I, Gabrielsson B (1983) Performance of musical rhythm in 3/4 and 6/8 meter. Scand J Psychol 24:193-213
Huber K (1923) Der Ausdruck musikalischer Elementarmotive. Eine experimentalpsychologische Untersuchung. Johann Ambrosius Barth, Leipzig
Jörgensen H (1988) Musikkopplevelsens psykologi [The psychology of music experience]. Norsk Musikforlag, Oslo
Kendall RA, Carterette EC (1990) The communication of musical expression. Music Percept 8:129-164
Langer SK (1953) Feeling and form. Routledge and Kegan Paul, London
Langer SK (1957) Philosophy in a new key, 3rd edn. Harvard University Press, Cambridge, Massachusetts (Original 1942)
Lindström E (1992) 5 x "Oh, my darling Clementine". The influence of expressive intention on music performance. Department of Psychology, Uppsala University
Nakamura T (1987) The communication of dynamics between musicians and listeners through musical performance. Percept Psychophys 41:525-533
Nielsen FV (1983) Oplevelse av musikalisk spaending [The experience of musical tension]. Akademisk Forlag, Copenhagen (with English summary)
Palmer C (1989) Mapping musical thought to musical performance. J Exp Psychol (Hum Percept) 15:331-346
Radocy RE, Boyle JD (1988) Psychological foundations of musical behavior, 2nd edn. Thomas, Springfield
Rigg MG (1964) The mood effects of music: a comparison of data from former investigations. J Psychol 58:427-438
Seashore CE (1938) Psychology of music. McGraw-Hill, New York (Reprint 1967 by Dover, New York)
Truslit A (1938) Gestaltung und Bewegung in der Musik. Wieweg, Berlin
Winner E (1982) Invented worlds. The psychology of the arts. Harvard University Press, Cambridge, Massachusetts

Expressive Aspects of Instrumental and Vocal Perfomance

JOHAN SUNDBERG, LARS FRYDÉN AND ANDERS FRIBERG

Introduction

Several music computers can now convert an input note file to a sounding performance. Listening to such performances demonstrates convincingly the significance of the musicians' contribution to music performance; when the music score is accurately replicated as nominally written, the music sounds dull and nagging. It is the musicians' contributions that make the performance interesting. In other words, by deviating slightly from what is nominally written in the music score, the musicians add expressivity to the music.

What are the deviations from the score, then? By means of measurements on actual performances physical correlates of musical expressivity have been identified in terms of numerous minor and major deviations from the pitches, durations, and sound levels nominally assigned to the different tones in the score (see, for example, Bengtsson 1975; Shaffer 1980; Todd 1985; Gabrielsson 1987; Clynes 1983; Repp this volume). These deviations are often termed expressive deviations although, for reasons to be explained below, the term "meaningful deviations" seems more appropriate.

Some years ago we started to investigate this phenomenon with the goal to gain a better understanding of the acoustic code used in music communication. The method was to improve the musical quality of computer-generated performances of input music files. In other words, we tried to educate the computer musically. We and the computer all learned a lot about music. Results from our research have been presented in detail elsewhere (Sundberg 1988; Sundberg et al. 1989, 1991a; Friberg 1991). In this article I will present some examples of meaningful deviations and their apparent purposes and discuss expressive aspects of musical performance.

Method

Our method is schematically illustrated in Fig. 1. The input is a music file in which pitches and durations are specified. This description is supplemented by information on major chord changes and phrase and subphrase boundaries. In the case of singing

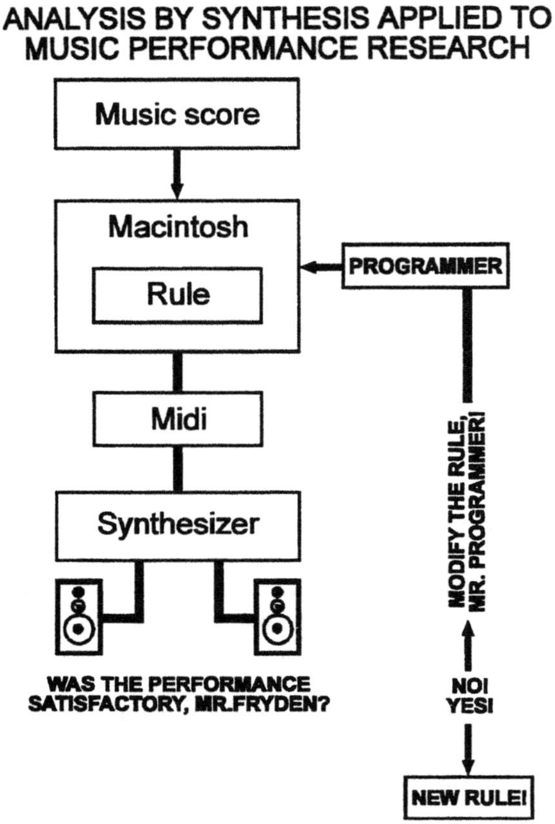

Fig. 1. How the analysis-by-synthesis procedure was applied to musical performance research

synthesis the text also is added in phonetic transcription, assigning phonemes to all tones.

The main work consists of formulating rules inducing meaningful variations in the performance. We call these *performance rules*. They have been assembled in a computer program written by Friberg (Fridberg and Sundberg 1986); Friberg 1991). This program automatically produces synthesized performances containing meaningful deviations.

The experimental procedure for tuning a rule is not all that different from a music lesson. A music teacher - Lars Frydén, string quartet violinist and conservatory music teacher - plays a decisive role. Normally he instructs players how to perform, listens to the result, and, if required, modifies his instructions. The difference in the computer application is that the student is a disastrous combination of complete obedience and lack of musical judgment. When used for synthesizing singing, some details are different. For instance, we have noted that many effects have to be somewhat greater in magnitude in synthesis of vocal music than in instrumental music. Also, the exact acoustic realization of consonants seems to affect expression.

In more technical terms, the music score is written into a microcomputer, which controls a synthesizer over MIDI. Meaningful deviations are generated by a set of ordered, context-dependent performance rules. These rules explicitly and exhaustively describe all the meaningful deviations that appear in the resulting computer-generated performances.

Examples of Rules

The rules as formulated affect duration, sound level, fine tuning, and the amplitude and frequency of the vibrato. Several rules affect the same parameter, so that, for example, a note can be shortened for many different reasons.

The rules can be divided into three groups. In western music, both pitch and duration intervals are categorically perceived (Burns and Ward 1978; Clarke 1987). By increasing the differences between pitch or duration categories, the *differentiation rules* seem to serve the purpose of helping the listener to recognize the pitch and duration categories. The *grouping rules* seem to serve the purpose of helping the listener to realize which tones belong together, i.e., to identify musical gestalts of various lengths. The *ensemble rules* concern synchronization and tuning within ensembles.

Some examples of rules are presented in the following. For a full technical specification the reader is referred elsewhere (Friberg 1991).

- *Duration contrast* shortens short notes and lengthens long notes. It thus increases the duration contrast between short and long notes.
- *High sharp* increases the fundamental frequency of notes: the higher the note, the sharper it is played.
- *Melodic charge* attempts to reflect quantitatively the fact that, given the harmonic and melodic context, some notes are more remarkable than others. The melodic charge of a scale tone is defined quantitatively using the circle of fifths with the root of the prevailing chord as the reference, as shown in Fig. 2a. The rule adds emphasis to the notes by increasing amplitude, duration, and vibrato depth, depending on the absolute value of their melodic charge, as illustrated in Fig. 2b. In addition it affects the fine tuning according to the melodic charge. Thus, scale tones on the right side of the circle of fifths are sharpened and those on the left side are flattened (see Fig. 2c).

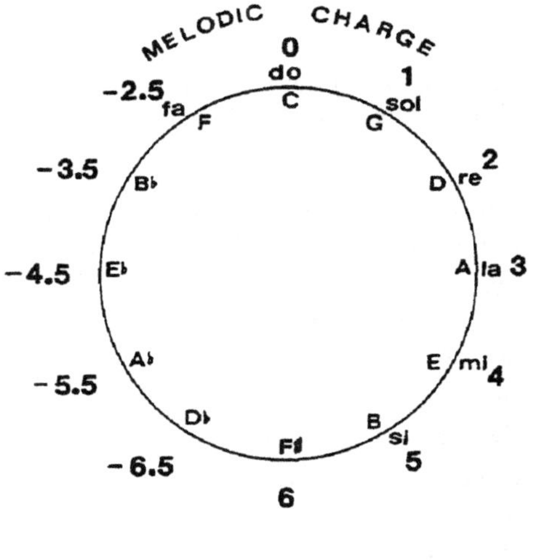

Fig. 2. a Definition, by means of the circle of fifths, of melodic charge in tonal music. The reference is the root of the prevailing chord.

Fig. 2. b Example of the effects of rule "melodic charge" on the theme of the first b minor Kyrie in the b *minor Mass* by J. S. Bach (BWV 232). The chords are given in the *top line* in terms of the interval in half steps between the root of the chord and the root of the tonic, minus sign denoting minor chords.

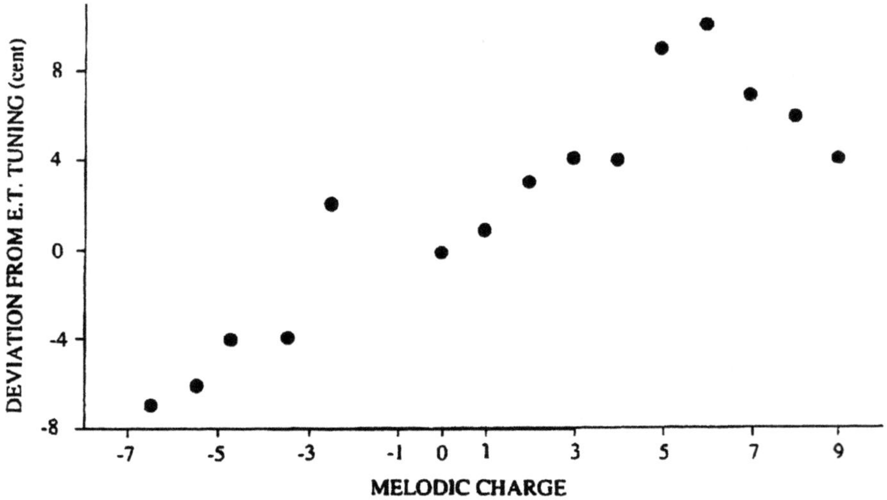

Fig. 2. c Deviations from the equally tempered tuning (*E. T.*) produced by rule "melodic intonation" which tunes the scale tones according to their melodic charge

- *Leap articulation* inserts a micropause between two tones forming a leap. The micropause is generated by letting the final part of the note´s amplitude decrease quickly before the onset of the following note. The duration of the pause, measured as the duration of the decay (DRO) (see Fig. 3) depends on the size of the leap up to nine half tone steps and remains constant for wider leaps. Figure 3 also shows how the rule affects a musical example.
- *Phrase marking* is a macro grouping rule operating on phrase and subphrase boundary signs which are added to the input notation. Figure 4 shows an example. It inserts an 80 ms micropause to the final note of subphrases and lengthens phrase final notes by 40 ms.
- *Harmonic charge* represents an attempt to reflect quantitatively the fact that in Western traditional tonal music, chords also form a hierarchical system reflecting

Table 1. Definition ofharmonic charge of some chords. The pitch names wihtin parentheses assume a C major tonality. Root distance specifies the number of semitones between the root of the chord and the root of the tonic.

	Harmonic charge	Root distance
Tonic (C E G)	0	0
Dominant (G B D)	2.0	7
Subdominant (F A C)	1.6	5
Double dominant (D F# A)	4.0	2
Double subdominant (Bb D F)	2.7	10
Relative of tonic (A C E)	1.3	9-
Relative of dominant (E G B)	5.0	4-
Relative of subdominant (D F A)	1.7	2-
Dominant of relative of tonic (E G# B)	6.5	4
Dominant of relative of dominant (B D# F#)	7.0	11
Dominant of relative of subdominant (A C# E)	5.4	9

their remarkableness. It equals a weighted sum of the chord notes' melodic charges computed with the root of the tonic as the reference (see Table 1). The macro grouping rule harmonic charge creates crescendos when a chord of higher *harmonic charge* is approaching and decrescendos in the opposite case, and these level variations are accompanied by proportional tempo and vibrato variations, as illustrated in Fig. 5b.

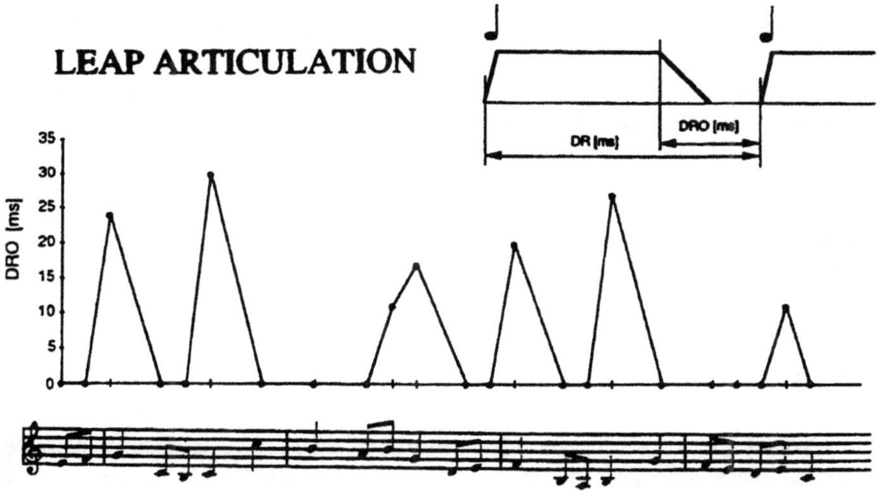

Fig. 3. Illustration of rule "leap articulation" which inserts a micropause, depending on the pitch separation between adjacent tones. The *ordinate* shows the duration of the decay (*DRO*) parameter, defined in the insert, and the *abseissa* is the serial position in the melody

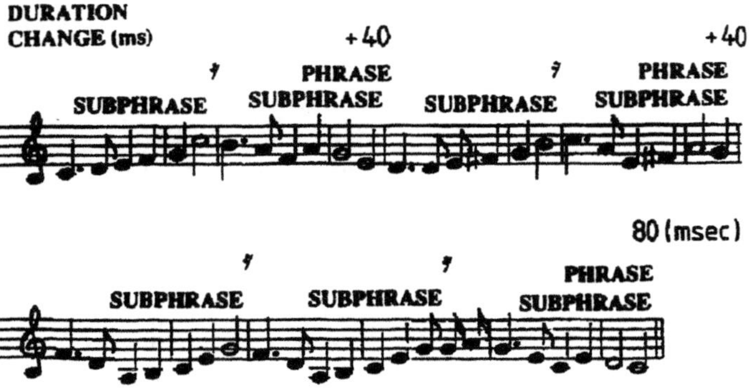

Fig. 4. Illustration of the effect of rule "phrase" which introduces subphrase markers in terms of an 80 ms micropause at the end of subphrases, and phrase markers by duration increases of 40 ms for phrase final notes; another 40 ms duration increase is made for the last note in the melody

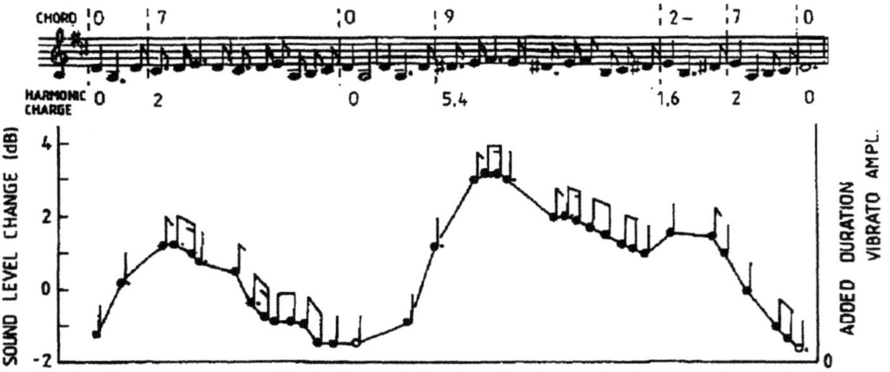

Fig. 5. Example of the effect of rule "harmonic charge" in terms of of the long-term changes in sound level, tempo, and vibrato. The chords are given in terms of the interval in semitones between the root of the chord and the root of the tonic

Verification

Listening Panels

One obvious way of testing the applicability of these rules is to have listening panels compare deadpan performances and rule generated performances and ask for preference (Thompson et al. 1989). However, we found this method difficult to use, because a single rule often produces effects which are so subtle that listeners often fail to note them. Furthermore, the rules do not seem to be independent; in other words, the musical effect of one rule depends on whether other rules are simultaneously applied.

Preferred Quantity

A rule possesses three important aspects with regard to its function: its *target* notes, i. e., the notes that it modifies, the *nature of modification* i.e., how it changes the performance, and the magnitude or *quantity* of the modification induced by the rule. In our performance program, the quantity of a rule can be varied by a factor Q, which is given the default value of 1.0 in the program.

Another way of testing rules is to ask musicians for preferred rule quantity. This was done in a production test for the rules mentioned above (Sundberg et al. 1991b). Six professional musicians were asked to adjust, by means of a slide ruler on the computer screen, the quantity of one rule at the time so that, in their judgment, the best performance of a music excerpt resulted. The available variation range was wide, included zero, and, if possible, also negative quantities.

Figure 6 shows averages and 95% confidence intervals for the preferred quantities. It can be seen that, with one exception, the preferred mean quantity for

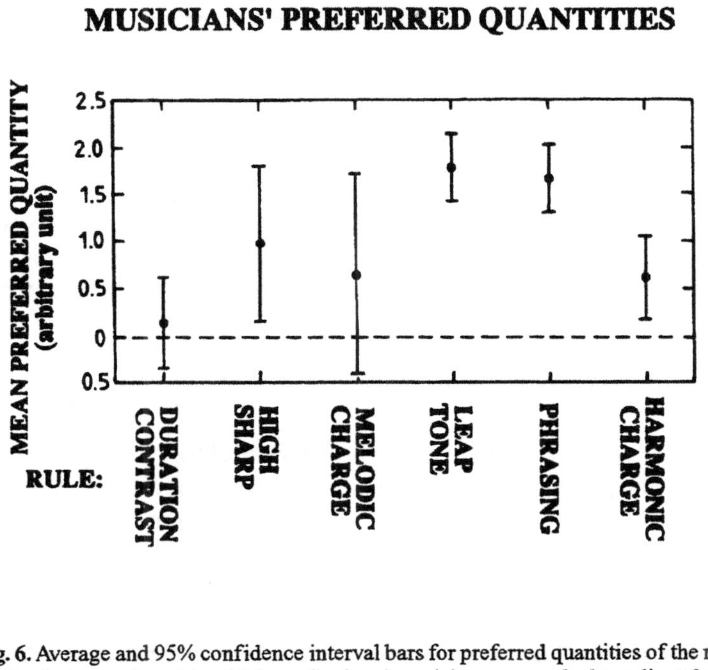

Fig. 6. Average and 95% confidence interval bars for preferred quantities of the rules tested in the production experiment in which professional musicians were asked to adjust the rule quantity to their own satisfaction. Quantity 1 is the default value in the performance program. In all cases but one (duration contrast) the musicians preferred a magnitude close to that produced by the rule system

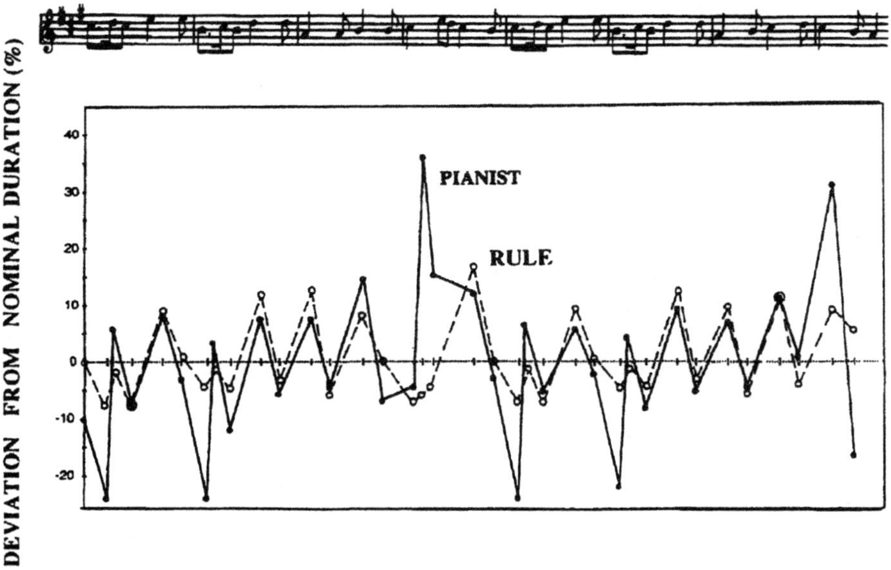

Fig. 7. Deviations from nominal note values performances of W. A. Mozart's *Piano Sonata in A major* (K 331). Observed data (*solid curve*) were taken from Palmer (personal communication). Predicted data (*dashed curve*) were obtained from the performance program using the default quantity parameter values. The rule "double duration" significantly contributed to the similarities between predicted and obeserved data, but it is also clear that some deviations occurring in reality are not produced by the rule system

all these rules was above zero. The exception was rule "*duration contrast,*" for which the preferred quantity was zero. In other words, the musicians did not find that the rule improved the performance. This was due to an unfortunate choice of excerpt (Sundberg et al. 1991b) or to other factors (Sundberg et al. forthcoming). For the other rules the result showed that they produced musically desirable effects.

Predicted and Observed Performance Data

Duration. Another method to test the musical relevance of the rules is to compare actual and predicted performances as produced by the performance program. Analyses of durations in music performances have been carried out by many authors (Sloboda 1983; Todd 1985; Clarke 1982, 1988). Figure 7 shows an example, a performance of Mozart's A major sonata for piano as measured by Palmer (personal communication, see also Palmer 1989). There is a good qualitative agreement although our rules rarely produce all meaningful deviations observed in this and other similar comparisons.

Fine Tuning. According to rule "*Melodic Intonation*" the fine tuning of the scale tones depends on the melodic charge. Thus, this rule predicts deviations from

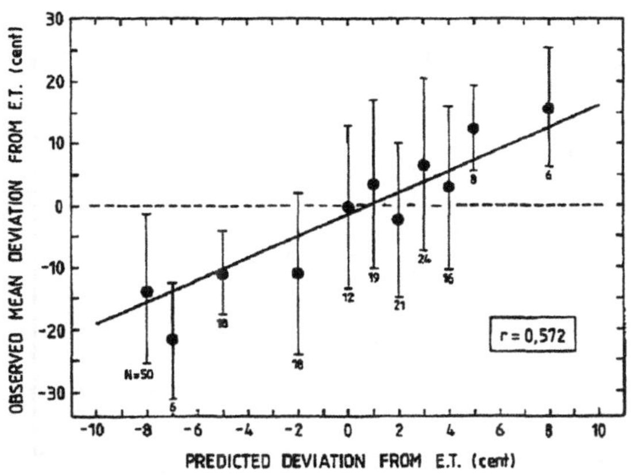

Fig. 8. Predicted and observed deviations from equally tempered tuning (*E.T.*) in performances of the solo part of the *Air* from J. S. Bach's *C major Suite* for orchestra. Performance data stem from Garbuzov (1948); performers are David Oistrach, Mischa Elman, and Efraim Zimbalista. Predictions were derived from rule "melodic charge" (see Fig. 2c). The *bars* represent ± 1 standard deviation and the *numbers* show the number of observations. The correlation shows that the deviations produced by this rule are qualitatively similar to those occurring in reality

the equally tempered tuning, depending on the melodic charge of the note.

Figure 8 compares such predictions with measure-ments by Garbuzov (1948) of performances of the first six measures of the solo part of the Air from J.S. Bach's C major Suite for orchestra played by David Oistrach, Mischa Elman, and Efraim Zimbalista. There is a significant correlation, indicating that the predictions show a qualitative agreement.

Independent Evidence

In order to attain appropriate performances we postulated two novel notions: melodic charge and harmonic charge. Independent evidence of these notions has emerged from experiments carried out by the psychologist Krumhansl and collaborators (see Krumhansl and Kessler 1982; Krumhansl et al. 1982). In these experiments, subjects judged how well a probe tone or probe chord served as a continuation of a preceding standard scale and standard cadence, respectively. Figure 9 shows the relation between these probe tone and probe chord ratings and melodic and harmonic charge, respectively.

There is a significant correlation in both cases. As one might expect, an excellent continuation (high rating) corresponds to a low melodic and harmonic charge, and vice versa. These results suggest that, while playing, musicians take into account listeners' ratings of how well the note or the chord served as a continuation.

Discussion

The experiments described above have shown that the musical quality of a performance can be raised if certain rules are applied that produce deviations from what is nominally written in the score. We have demonstrated that our rules produce deviations that occur in real performances. Thus we can conclude that our rules

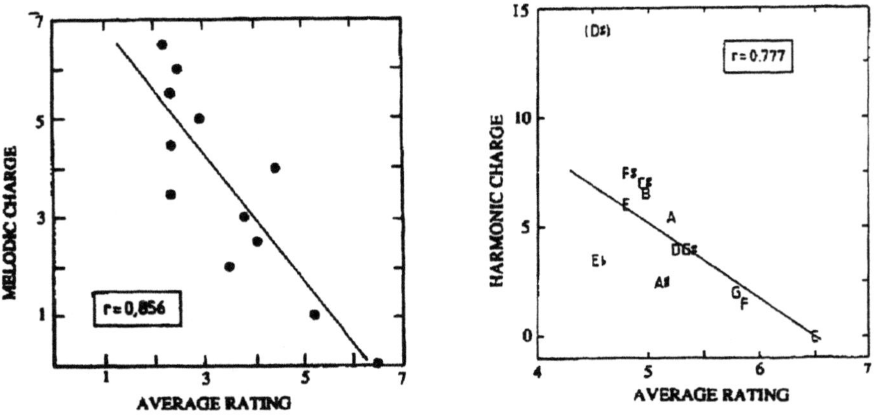

Fig. 9. Melodic and harmonic charge are entities, the existence of which we postulated when we developed the performance rules. They show a significant correlation with results from independent research made by Carol Krumhansl and collaborators in which subjects rated how well a probe tone or chord served as a continuation of a standard scale and a standard cadence, respectively. The correlations suggest that tones and chords that appear surprising to a listener have a high melodic and harmonic charge, i. e., they are emphasized in the performance

induce deviations that create musical expressivity. We have also seen that the concepts we needed to construct for some of the rules, namely, melodic and harmonic charge, correlate with basic functions in music listening as revealed by completely different methods.

Apparently, most of our rules serve either of two main purposes, namely, to facilitate the listener's differentiation and grouping of the sounds, while some other rules concern technicalities such as ensemble tuning and common timing. Note that we do not make any claims as to the generality of the rules; some rules probably have a general applicability while others are optional or depend on the instrument or the musical style. However, aspects of musical expression and communication seem more interesting. For instance, we may speculate around questions such as what exactly it is that the performer expresses by means of the differentiation and grouping rules? What information is transferred to the performer by means of musical expressivity?

The differentiation rules appear to aim at making the identification of pitch and duration categories easier for the listener. This is realized by enhancing the differences between pitch and duration categories. This enhancement seems to use two methods. One is to simply increase the physical separation between the categories. Examples are "*duration contrast*" and "*high sharp*" which shorten short notes and sharpen high notes, respectively. Another method is to add a difference between the tones in a secondary property. For example, rule "*melodic charge*" adds emphasis to an extent that reflects the melodic charge. This emphasis is marked by means of several secondary properties: sound level, duration, vibrato depth, and fine tuning.

The information conveyed by means of the grouping rules obviously concerns musical structure. An obvious example of a grouping rule is "*phrasing*," inserting a micropause at the end of subphrases and lengthening the final note of phrases. "*Leap articulation*" inserts a micropause between tones forming a melodic leap thus suggesting that the leap is a sign of a structural boundary. It should be noted, though, that the composer may indicate violation of this rule by inserting a slur sign across the leap. "*Harmonic charge*" also seems to induce a grouping effect by producing crescendos and decrescendos; the continuous increase or decrease of loudness suggests that the tones involved in these changes belong together.

Summarizing, it seems fair to conclude that the differentiation and grouping rules serve the purpose of enhancing the musical structure. In other words, one function of musical expression is to assist the listener in understanding the musical structure, i.e., to facilitate the processing of the flow of acoustic signals produced.

Traces of the same propensity to facilitate the listener's differentiation and grouping of sounds can observed in areas other than music. Speech sounds are also categorically perceived and a speaker typically enhances the difference between the categories (Carlson et al. 1989). For example, short vowels are not only pronounced with shorter durations than long vowels, they also have a different vowel timbre in many languages. Furthermore, grouping is marked by pauses and by many other means.

It is interesting to observe that the idea to facilitate differentiation can be found also outside the area of acoustic communication. For example, architects often choose one color for all windows, another color for the walls, and a third for the roof. Thus, also in other communication systems it seems important to facilitate differentiation.

By marking remarkableness, quantified in our program in terms of melodic and harmonic charge, the musician takes the hierarchically organized system of scale tones and chords into account. The means thereby used seems to be *emphasis*: melodically charged notes sound emphasized in our computer generated performances. Emphasis is related to *predictability*. By emphasis the listener is alerted so that communication may be maintained even at turns in the signal flow which the listener is not likely to predict. The same need for emphasizing unexpected elements can also be observed in speech. Predictability seems to be another important component in a communication system.

As our performance rules define how the meaningful deviations are effected acoustically, they provide information as to the means used. Thus we may ask what code is used for transferring the information conveyed by these deviations. This will enable us to find out if this code is unique for music. Acquaintance with this code must be relevant to musicality; if a listener does not understand this code, much of the essence of music listening may remain hidden to him.

According to the "*phrasing*" rule, phrase endings are marked by lengthening the final note of phrases. Again, we find striking parallels with speech, where lengthening of syllables is typically used for marking the ending of sentences and smaller structural units (Carlson et. al. 1989; Todd 1985).

The rule "*melodic charge*" adds duration and loudness to tones according to their melodic charge. The perceptual effect of this is that notes with high melodic charge sound emphasized. The same applies to speech where emphasis is typically marked by increased syllable duration and loudness (Carlson et al. 1989; Sundberg et al. 1991a).

A tempting conclusion would be that speech is the origin of the code. However, the code used in speech varies to some extent between languages while musicianship seems largely independent of the player's mother tongue. This suggests that the origin of the code used in musical performance originates elsewhere.

One possible source of code is the functioning of the perceptual system. It seems reasonable to assume that this system needs slightly more time for processing unexpected information than expected information. A similar need for extra processing time seems to exist at the end of structural constituents such as phrases and is manifested both in speech prosody and music composition (Lindblom 1978).

Another source of the acoustic code used for the meaningful deviations in music performance is the idea of icons. Some of the performance rules seem clearly iconic. The rule "*harmonic charge*" produces crescendos leading up to a new, harmonically more charged chord or a decrescendo to a harmonically less charged chord. The underlying purpose would be to keep all the notes together. This may be accomplished by letting all those satellite notes participate in one and the same long-term event, namely, an increase or decrease of sound level.

Icons may of course be selected from many different fields. One such field of some significance seems to be locomotion. The average pattern of ritardando revealed by measurements on 24 recordings of music from the Baroque or Classical eras is shown in Fig. 10 (Sundberg and Verrillo 1980). It was found to be nearly identical with the curve showing the decrease of footstep rate during stopping of locomotion in running, provided that neither step length nor braking force was changed during the stopping process, i.e., that the stopping was perfectly preplanned (Kronman and Sundberg 1987). This supports the assumption that listeners under

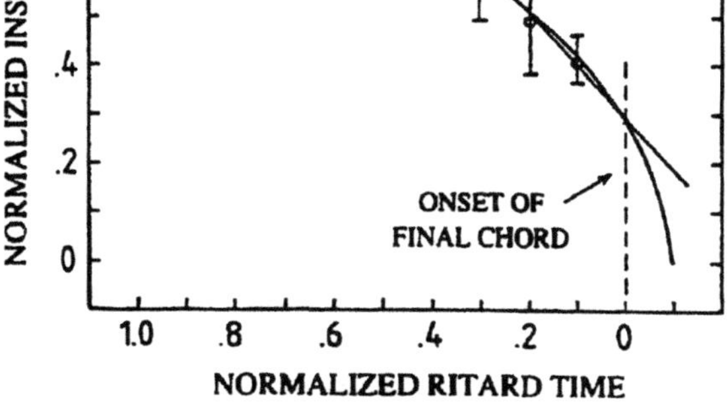

Fig. 10. Mean slowing of the tempo, i. e., the inverse of the beat duration during final ritardandos of 24 pieces for keyboard instrument from the baroque era (*circles*); the *bars* represent ± 1 standard deviation. The *solid line* indicates the inverse of the time intervals between footsteps during an idealized stopping of running. The similarity between the *curves* suggest that the final ritardando alludes to the listener's experience of a perfect stopping of running, thus providing the meaning of the final ritardando: "Here comes the end of the piece"

stand the meaning of a ritardando because it is iconic: it alludes to the listener's experience of stopping locomotion in running.

We can conclude that the code used for the meaningful deviations in musical performance does not seem specific to music. This implies that music is not an exclusive form of art, demanding special knowledge. On the contrary, music is available to almost anyone, as can also be expected, since very nearly 100% of the population enjoys listening to music, though to various types.

Conclusions

What do musicians add to the music described in the score, and why do they add this? Our analysis-by-synthesis of music performance allows us to propose hypothetical answers to these questions.

Musicians add and subtract duration, sound level of the tones, and also vary fine tuning and vibrato characteristics, if possible. By these means musicians help the listener to correctly categorize the sound events, to correctly identify structural boundaries, and to raise the listener's attention when remarkable events occur. Striking similarities between musical performance and speech have been revealed. Speakers also use prosody to assist the listeners in correctly categorizing the speech signals, to group them, and to add emphasis on remarkable events. There are also striking similarities regarding the acoustic means used for conveying this information; shortening and lengthening seem to be used with the same meaning, and the means to signal emphasis are largely identical.

Acknowledgments. The author is indebted to Dr. Jaan Ross, Tallinn, Estonia, for showing him the cited data published by Garbuzov. Part of the research was supported by the Tercentenary Foundation of the National Bank of Sweden. This is a revised and partially rewritten version of a paper presented at the FASE meeting in Zürich, July 1992.

References

Bengtsson I (1975) Empirische Rhythmusforschung in Uppsala, (Hamburger Jahrbuch für Musikwissenschaft vol. 1) Wagner, Hamburg: pp 195-220

Bock RD (1975) Multivariate statistical methdos in behavioral research. McGraw-Hill, New York

Burns E, Ward D (1978) Categorical perception, phenomenon or epiphenomenon: evidence from experiments in the perception of melodic musical intervals. J Acoust Soc Am, 673:456-468

Carlson R, Friberg A, Frydén L, Granström B, Sundberg J (1989) Speech and music performance: parallels and contrasts. Contemp Music Rev 4:389-402

Clarke E (1982) Timings in the performance of Erik Satie's 'Vexations'. Acta Psychol (Amst) 50:1-19

Clarke E (1987) Categorical rhythm perception: an ecological perspective. In: Gabrielsson A (ed) Action and perception in rhythm and music. Royal Swedish Academy of Music, Stockholm, pp 19-33 (Publ no 55)

Clarke E (1988) Generative principles in music performance. In: Sloboda J (ed) Generative processes in music. Clarendon,.Oxford, pp 1-26

Clynes M (1983) Expressive microstructure in music, linked to living qualities In: Sundberg J, (ed) Studies of music performance. Royal Swedish Academy of Music, Stockholm, pp 76-181 (publ no 39)

Friberg A (1991) Generative rules for music performance. A formal description of a rule system. Comput Music J
Friberg A, Sundberg J (1986) A Lisp environment for creating and applying rules for musical performance. In: Berger P (ed) Proceedings of the international computer music conference, The Haag: Royal Conservatory, 1-3
Gabrielsson A (1987) Once again: the theme from Mozart's piano sonata in A major (K. 331). In: Gabrielsson A (ed) Action and perception in rhythm and music. Royal Swedish Academy of Music, Stockholm, pp 81-104 (publ no 55)
Garbuzov N (1948) Zonnaja priroda zvukovysotnogo slucha. Moscow: Akademija Nauk SSSR, Moscow
Kronman U, Sundberg J (1987) Is the musical ritard an allusion to physical motion? In:. Gabrielsson A (ed) Action and perception in rhythm and music. Royal Swedish Academy of Music, Stockholm, pp 67-68 (Publ no 55).
Krumhansl C, Kessler E (1982) Tracing the dynamic changes in perceived tonal organisation in the spatial representation of musical keys, Psychol Rev 89: 334-368..
Krumhansl C, Bharucha J, Kessler E (1982) Perceived harmonic structure of chords in three related musical keys, J Exp Psychol (Hum Percept), 8:24-36
Lindblom B (1978) Final lengthening in speech and music, In: Gårding E, Bruce G, Bannert R (eds): Nordic prosody. Department of Linguistics, Lund University pp 85-101
Palmer C (1989) Mapping musical thought to musical performance. J Exp Psych (Hum Percept 15; 331-346..
Shaffer L (1980) Analysing piano performance: a study of concert pianists. In: Tutorials in motor behavior. Stelmach G, Requin J (eds) North-Holland, Amsterdam, pp 443-455
Sloboda J (1983) The communication of musical metre in piano performance. Q J Exp Psych (A) 35:377-396..
Sundberg J (1988) Computer synthesis of music performance.In: Sloboda J (ed.) Generative processes in music. Clarendon, Oxford, pp 52-69
Sundberg J, Friberg A, Frydén L (1989) Rules for automated performance of ensemble music. Contemp Music Rev 3:89-109
Sundberg J, Friberg A, Frydén L (1991a) Common secrets of musicians and listeners. An analysis-by-synthesis study of music performance. In: Howell P, West R Cross I (eds) Representing musical structure. Academic, London
Sundberg J, Friberg A, Frydén L (1991b) Threshold and preference quantities of rules for music performance. Music Percept 9:71-92
Sundberg J, Iwarsson J, Hagegard H (forthcoming) A singer's expression of emotions in sung performance. In: Fujemura O (ed) Vocal Fold Physiology, San Diego: Singular Press
Sundberg J, Verrillo V (1980) On the anatomy of the ritard: a study of timing in music. J Acoust Soc Am 68:772-779
Thompson W, Sundberg J, Frydén L (1989) The use of rules for expression in performance of music. Psychol Music 17:63-82
Todd N (1985) A model for expressive timing in tonal music. Music Percept 3:33-58

The Two Steps of the Categorization Process in Music Listening
An Approach of the Cue Extraction Mechanism as Modular System

Irène Deliège

Introduction

This chapter concerns mainly the categorization processes in music perception within the framework of a hypothesis of cue extraction during listening which I developed some years ago (Deliège 1987, 1989, 1991). This concept will be briefly explained before turning questions relating directly to categorization starting from cue extraction. The questions will be applied to three categories of subjects: adult subjects, children from 9 to 12 years of age and one autistic child of the same age. This three-tiered approach was adopted in an attempt to show that categorization processes in music listening are comprised of two distinct steps: the first - cue extraction - is peripheral and perceptual, here suggested to be a modular system, while the second - access to the evaluation of items during listening - is central. This research was sparked by an assertion Neisser (1987) wrote some years ago: "the processes of categorization intervene rather in a second phase of perception: categorization is somewhat less 'direct' than perception. The perceptible properties of things are specified by invariants to which the perceptual systems need only resonate" (p.22). It would appear, then, that the two aspects should not be confused, but considered as distinct aspects of the same general functioning.

It has long been held that the organization of groupings is the key to a good memory. But I would like to go further by suggesting that the mastering of music listening is organized on the basis of the extraction of salient cues selected from within the groups themselves. Some structures are thus "marked" and coded selectively in such a way that they may thereafter serve as reference points. The role of cues is to generate abbreviations of the units created which reduce the amount of information to be stored. Cues contain the invariants of discourse, thereby remaking it possible to continuously evaluate new entries. Thus, cues play an integral role in the categorization process. They are first extracted from the diversity of sounds perceivable in a work and create their own "auditory images" (Leipp 1977) independent of each other. Therefore, they permit structures to be identified and become a reference for the processes of classification and comparison.

Categorization in Adults

Categorization is generally discussed in terms of semantic and conceptual representation. Rosch (1975, 1976) defined two dimensions: verticality and horizontality. Verticality establishes the relations between categories, which leads to the idea of a hierarchy between the various category levels. As we know, Rosch proposes three levels: the superordinate level, the basic level, and the subordinate level. The basic level in the categorization of listening to music may be at that point where different cues are extracted within the same work. They each have their own function and create their own particular auditive image. The superordinate level may then be conceived in the reference that each cue defines to a grouping within the mental representation of the work.

The subordinate level and the notion of horizontality, in some sense, intersect. This notion must be understood as the gradual spacing out of different items within an ensemble. It is immediately clear that the perception of music will exploit this principle of horizontality to the utmost, because it means that all the possible variations of a basic cell - each cue - engender their own relationships of horizontality.

In relation with this idea, I would like to present an overview of three experiments I have carried out recently, taking care to limit myself to a discussion of those results which back up my argument. These studies all follow the same basic structure: (1) evaluating the frequency of appearance of two motifs in a piece of music; (2) classifying the motifs into their respective families; (3) and evaluating the degree of similarity between pairs of motifs.

Pieces constructed on two main motifs were selected deliberately. Subjects were first to listen to the complete piece once. Then, after listening to the two main motifs, they were requested to assess their frequency of appearance in the piece. They were then to classify all the motifs of the piece into their respective families. Finally, the subjects were asked to evaluate the degree of similarity between pairs of motifs. Each pair contained the reference motif as well as one of its derived motifs. Some "decoys" were incorporated into this ensemble: pairs having total similarity (i.e., the repetition of two identical motifs) and pairs with no similarity (i.e., motifs from different families).

The piece selected for the experiment with adult subjects (25 nonmusicians and 12 professional musicians) was the Finale of the Sonata for Solo Violin in C major by J.S. Bach (Fig.1; Deliège 1990-91).

The frequency of appearance of the reference motifs was to be evaluated on a scale from 1 to 3. Musicians recorded average results of 2 for motif A and 3 for motif B. Nonmusicians' responses were less clear-cut. Motifs A and B received average scores of 2.16 and 2.52, respectively.

They then memorized reference motifs A and B (Fig.2) over eight consecutive listenings before classifying the derived motifs (10 A motifs and 14 B motifs). The motifs were heard three times in random order without interruption. There were thus 72 sequences.

No errors were recorded for the musicians, while the nonmusicians had an average of 94.5% correct responses.

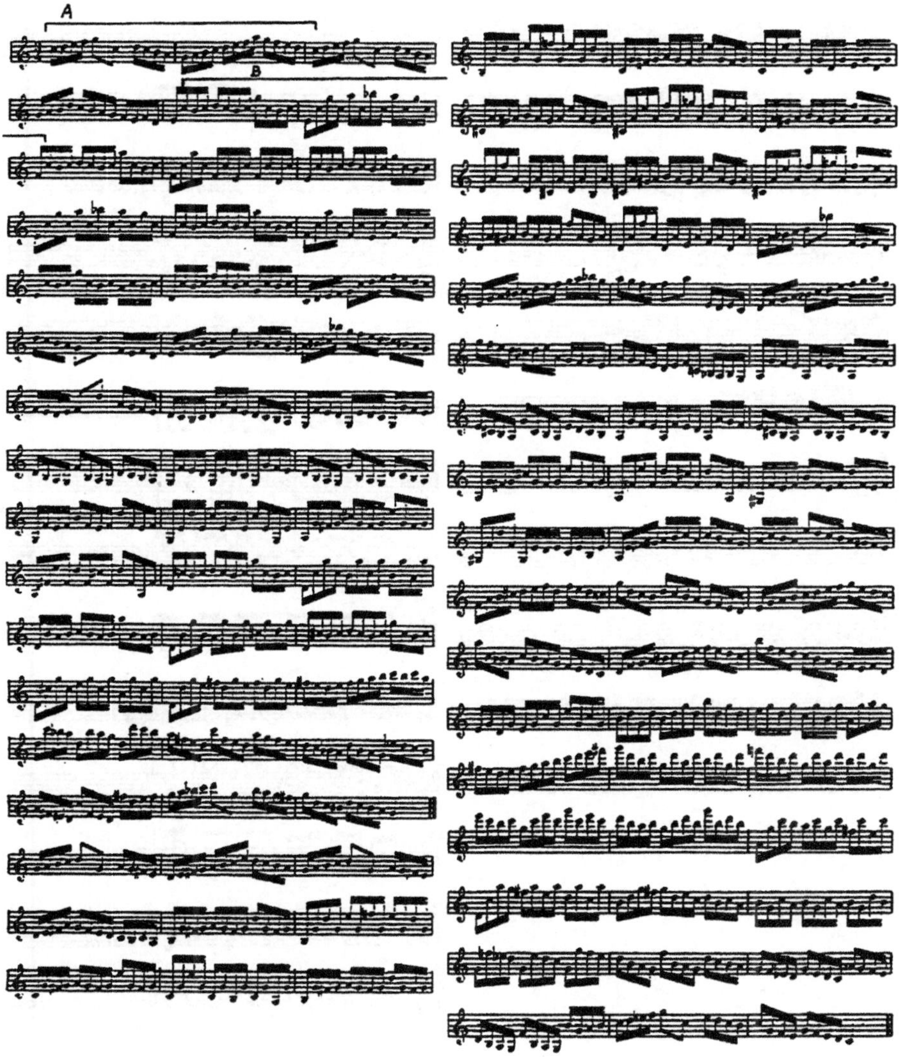

Fig. 1. Finale of the Sonata for Solo Violin in C major by J.S. Bach. The reference motifs *A* and *B* are indicated by a *bar*

Next, the degree of similarity of pairs of items were evaluated on a scale of seven (0 = no similarity to 6 = complete similarity). Subjects were not instructed to wait for the end of the item before replying, this in an attempt to determine whether they replied when in possession of partial elements of the information available. The test used the same sequences plus the four "decoys." The items were heard three times. The first item was always either reference motif A or B followed by one of their derivations for comparison purposes. There were thus 76 sequences of two items.

Fig. 2. Reference motifs *A* and *B* and all derived motifs (10 A motifs and 14 B motifs)

The results show that family B was felt to be more homogeneous: very similar results are observed in both categories of subjects. On the other hand, family A contains examples which are more distant from the reference motif (Fig.3)

The "decoys" were employed to measure the degree of persistence in listening. Indeed, in all other pairs of items there is a certain degree of variation from the reference motif. None of the musicians were in error here. In nonmusicians, total similarity responses were given when the items began with a group of identical sounds (mainly in items 1, 10, 11 and 14) and when degrees of zero similarity were recorded (for items 2, 4 and 8) for the opposite reason. The nonmusicians' attention was thus less efficient, their responses being given on the basis of an initial fragment only.

Categorization in Normal Children

The second part of the chapter will deal with the way in which categorization processes in listening to music are defined in normal children. Are the processes already observed in adults present here as well ?

Specialists in conceptual development in children assert that although children possess taxonomic competency at a rather early stage, this ability is probably overshadowed by other types of categorical functioning more closely related to their everyday lives. So, although children are capable of grouping items according to taxonomic characteristics when they believe they are learning a novel word (Markman and Hutchinson 1984), categorization naturally develops on a thematic or functional base (Fivush 1987; Markman 1987) more directly linked to their representation or scheme of events (in simpler terms their daily experiences). The child will thus readily group together objects in terms of their being part of a common routine (i.e., a thematic category as for example pyjamas and the bed, a spider and his web) or of their being able to be substituted for one another at the same time (i.e., a functional category - at breakfast everything which can be put on bread: butter, cheese, jam, etc).

Fig. 4. Ländler D 145, op. 18, no. 10 by F. Schubert. The reference motifs A (called "Pierre") and B (called "Anne") are indicated by a *bar*

As far as music is concerned, given the absence of a semantic referent, it is hardly likely that interference due to the child's tendency to prefer categories in relation with his daily life interferes with categorization. It is more likely that children will act as they do in new word acquisition and strongly rely on typicality features, i.e., the cues extracted while listening. In this sense, the categorical level, as in adults, is defined as the basic level (Rosch, et al. 1976 ; Mervis 1987). We may thereby assume that children are able to perform thusly, since when Rosch's principles of categorization are extrapolated to listening to music, cue extraction is situated precisely at this basic level. Children's performances probably extend even further, i.e., to the comparison of motifs derived from the same reference motif.

An experiment was conducted on normal children from 9 to 12 years of age, one group with 2 or 3 years musical training and the other with no formal musical background (Deliège and Dupont 1994). The piece of music chosen was shorter so as to better suit the age of the subjects: a *Ländler* by Schubert (Fig.4). The two main motifs were called Pierre and Anne, to make the task less abstract for the children. You will note that motif A ("Pierre") was heard eight times in the piece and motif B ("Anne") only four times.

Summary of Results

For the frequency evaluation of the appearance of the two motifs, the question was "Did you hear "Pierre" or "Anne" more often?" Results in children with a musical background were very good (72% for motif A and 28% for motif B) while those of the nonmusicians were poor (41% for motif A and 59% for motif B). One may thus suppose that this ability requires a certain degree of development and knowledge.

As expected, the classification of the items derived from reference motifs A and B was well done: 79% correct responses were given by musician children and 70% by non-musician children. They thus reproduce with music sequences the same taxonomic competence as in new word acquisition (see above) and show a strong ability in cue extraction.

The evaluations of similarity were given on a scale of 0 (no similarity) to 4 (total similarity): it is thus less demanding than that presented to the adult subjects (Fig.5). The results are very close for the two groups of children; except for the first motif, they are located for the most part in the middle of the scale. Furthermore, the evaluations of the musician children are slightly lower. However, it is impossible to attribute this to their musical knowledge.

As for the appreciation of absence or total similarity, the musician children posted better scores: 86% correct responses were given for items having no similarity and 79% for total similarity, while the nonmusicians got 70% and 63% respectively. Again, the contribution of musical training is clear here: it accounts for the relatively rapid development of a particular sort of attention in children.

In summary, children clearly have access to cue extraction while listening to musical stimuli (or the basic level to use Rosch's terminology). Moreover, we know

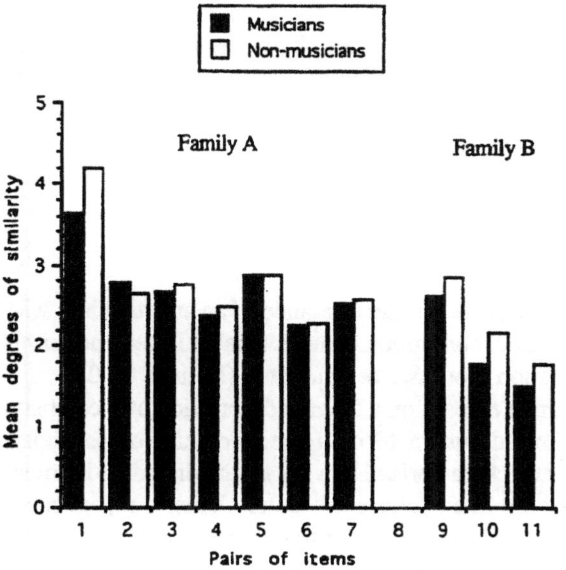

Fig. 5. Average degrees of similarity ratings for all pairs of motifs *A* and *B* of the Ländler D *145, op. 18*, no. 10 by F. Schubert

that this aptitude is present from a very early age. Research by Trehub et al. 1987) has shown that infants are capable of encoding characteristics which enable them to recognize a melodic pattern even after transposition and modifications in tempo (Trehub and Thorpe 1989) i.e., when the cues extracted remain largely intact. Conversely, changes in the melodic curve or in rhythmic organization adversely affect recognition in infants (i.e., whenever musical transformations interfere with extracted cues). This last observation in particular is of special importance.

Table 1. General overview of the results of the three experimental approaches

	Classification		Frequency		Similarity		No or total similarity	
	Mus.	N.Mus.	Mus.	N.Mus.	Mus.	N.Mus.	Mus.	N.Mus.
Adults	Good +	Good	Good +	Good	Good +	Good	Good +	Good
Children	Good	Good	Good +	Bad	Good	Good	Good	Good -
Autistic child	Good		Bad		Impossible		Impossible	

Mus., Musican; N.Mus., Non-musican

On the other hand, we have seen that the evaluations of the frequency with which motifs appear and of the degrees of similarity between musical structures seem to depend more on training and knowledge. Two distinct steps would thus seem to underlie the processes used in the tasks described earlier. (Table 1 summarizes what has been discussed up until this point).

Categorization in an Autistic Child

The third part of the chapter will examine the performance of an autistic child who was subjected to similar experimental conditions as the normal children and how it may shed light on the categorization processes at issue here (Heinen 1992).

It is certainly not my intention to embark on a lengthy description of the autistic syndrome. In 1943, Kanner showed autism to be a specific form of psychosis declaring itself early in infancy and characterized by very severe disorders in social and cognitive development. I would like to stress this latter element in particular. In *Autism: Explaining the Enigma*, Frith (1989) demonstrates the existence in autistic children of a rupture between early cognitive processes (i.e., peripheral perceptual processes) and central processes. She stresses the fact that only the latter are affected: "the problem of autism is not a peripheral problem; it involves the mental and not the sensorial apparatus" (p. 174). Uta Frith's assertion is not unique; Hermelin and O'Connor (1970), Peeters (1989), Leboyer (1985, pp.102-109) and Lovaas et al. (1979), among others, point out that the autistic child is capable of making subtle and detailed distinctions but not able to generalize on the basis of a piece of acquired knowledge and give it meaning. In other words, he can gather up characteristics thanks to a sort of perceptual reflex without being able to attain the concept itself.

On the basis of these premises, it should be possible to make more precise observations concerning normal children: the dissociation conspicuous between peripheral and central processes should make it possible to more clearly determine the expression of the two steps in the process.

While the young autistic girl failed in the evaluation of the frequency of motif appearance, she had little difficulty in classifying derived motifs in relation to a reference motif. Her score of 80% correct responses proves that the invariant cues had indeed been extracted. In the tests evaluating degrees of similarity between pairs of motifs, she responded repeatedly in terms of these invariants, thereby reproducing a classification process. When presented with a pair of motifs, she never waited for both items to be completed, but gave an almost immediate response for each, sometimes even before it was finished. She said either "It's Pierre" or "It's Anne," answers which were almost always correct but which indicate that the processes of comparison and evaluation were not established (see Table 1).

In conclusion, the autistic subject's performances clearly show the two stages of a completed categorization process in music listening. The first, peripheral in nature (in this case cue extraction), is perfectly operative and distinct from the second which is beyond the reach of the autistic child. Moreover, I stressed earlier the presence of this first step in infants on the basis of results obtained by Trehub et al.(1985,1987).

And If Fodor Was Right ?

Could not all these data have implications for Fodor's (1983) thesis of modularity? Could the process of cue extraction belong to the modular architecture of the human mind? Fodor first postulates that all modules are by definition peripheral, a general condition which is now fulfilled. He then considers the set of properties that a modular system must incorporate to as great a degree as possible. First is its specificity to a particular sensorial modality (first property): in other words, a module particular to hearing is not functional in relation to smell etc. Consequently, a module is encapsulated (fifth property), i.e., indifferent to what is operative in surrounding systems and apt to undergo specific traumatisms (eighth property) since it is bound to a designated neuronal organization (seventh property). A modular system operates automatically (second property) and may not escape from an operation which is by nature very rapid (fourth property). The central system has only limited access to the internal computational content of a modular system (third property). Conversely, the central system, as long as the data supplied by all modules are relatively superficial, has little difficulty in interpreting outgoing content (sixth property). Finally, modular systems are "prewired": their development is triggered by contact with the environment while the type of information likely to concern a particular modular system is determined innately (nineth and last property).

How many of these properties then are manifest in our subjects´ performances? The extraction of cues from a musical message is not only part of the auditive modality but is particular to music listening (first property). That a module of this type is self-contained and perfectly autonomous is a logical postulate indeed (fifth property): in other words, its faculties would not be shared by surrounding systems (e.g., language). Fodor gives as example the fact that the judgment of musical ideas has nothing in common with the judgment particular to mathematical ideas. In addition, the gathering together of cues operates in an automatic and mandatory fashion (second property) and the central systems have easy access to the module's content (sixth property) for those evaluative operations necessary for active music listening. The nineth property - the system's coming into existence at a relatively early point - appears in the interpretation of Trehub's results on the basis of the hypothesis of cue extraction put forth earlier. Finally, the speed at which recognition takes place (fourth property) has not entering into the experiments discussed until now. But results of a study conducted recently on the recognition of popular songs following transformations of the melody by reduction or relatively drastic shifts in the segmentations of the rhythmic group (Mélen and Deliège, in press) give the expected support needed at this point. A particular methodological disposition measured the speed of response. It was found that speed of recognition was intimately related to the relative amount of change the original cues were subjected to: the more damaged the cues were, the slower recognition was.

In conclusion, six of the properties mentioned earlier were functional in the results that have been proposed. Properties 3, 7 and 8 were not at work in the data presented. In fact, items 7 and 8 fall within the domain of neuropsychology, an area not touched on here. With the reservation that such approaches may contradict what has just been suggested, the modular aspects of the process of cue extraction in music listening offers substantial advantages. In addition, a system does not necessarily have to respond to all the properties enumerated earlier to be considered as modular.

Consequently, the idea of a hierarchy within the architecture of the modular system itself is indeed justified. By definition, a cue may contain different musical parameters. Peretz and Morais (1989) have recently proposed the encoding of pitch as well as rhythmic organization as modular. Cue extraction would then have to be situated at a higher level since it can include them both due to its variable number of parameters. This proposal is close to the idea put forth in language processing of a subdivision of the specialized modules into submodules. (Segui and Beauvillain 1988 ; Segui 1992; Coltheart and Davies 1992).

Conversely, the possible interference of an underlying harmonic structure which proved to be slower to access in the recognition processes of melodic structures would probably not reveal (at least not exclusively) a modular system, but central system instead. Modularity is only at work in situations close to reflex at early and local steps of information processing. To take into account a harmonic structure implies relations to contextual information which expand over more or less long periods of time. It follows that levels of representation containing these aspects would be likely treated by central systems.

References

Coltheart M, Davies M (1992) Le concept de modularité à l'épreuve de la neuropsychologie. In: Andler D (ed) Introduction aux sciences cognitives. Paris, Gallimard, pp. 109-130

Deliège I (1987) Le parallélisme, support d'une analyse auditive de la musique: vers un modèle des parcours cognitifs de l'information musicale. Anal Musicale 6, 73-79

Deliège I (1989). A perceptual approach to contemporary musical forms. Contemp music Rev 4:213-230

Deliège I (1991) L'organisation psychologique de l'écoute de la musique. Des marques de sédimentation - indice, empreinte - dans la représentation mentale de l'oeuvre. PhD thesis, University of Liège

DeliègeI, Dupont M (1994) Extraction d´indices et catégorisations dans l'écoute de la musique chez l'enfant. In: Deliège I (ed) Proceedings of the 3rd ICMPC. Escom, CRMW, Lièg

Fivush R (1987) Scripts and categories. In: Neisser U (ed) Concepts and conceptual development. Ecological and intellectual factors in categorization. Cambridge, Cambridge University Press, pp 234-254

Fodor JA (1983) The modularity of mind. MIT Press, Cambridge Mass

Frith U. (1989) Autism: explaining the enigma. Blackwell, Oxford

Heinen J (1992) La perception musicale chez l'enfant autiste. Une étude de cas unique. Masters-thesis, University of Liège

Hermelin B, O'Connor N (1970) Psychological experiments with autistic children. Pergamon, Oxford
Kanner L (1943) Autistic disturbances of affective contact. Nerv.Child 2:217-250
Leboyer M (1985) Autisme infantile. Faits et modèles. Presses Universitaires de France, Paris
Leipp E (1977) La machine à écouter. Masson, Paris
Lovaas I, Koegel R, Schreibman C (1979) Stimulus overselectivity in autism: a review of research. Psychol Bull 86(6):1236-1254
Markman EM, Hutchinson JE (1984) Children's sensitivity to constraints on word meaning: taxonomic vs thematic relations. Cogn Psychol 16:1-27
Markman EM (1987) How children constrain the possible meanings of words. In: Neisser U (ed) Concepts and conceptual development. Ecological and intellectual factors in categorization. Cambridge University Press, Cambridge pp 255-287
Trehub SE, Thorpe, LA, Morongiello BA (1985). Infants' perception of melodies: changes in a single tone. Infant Behav Dev 8:213-223
Trehub SE Thorpe LA, Morongiello BA (1987) Organizational processes in infants' perception of auditory patterns. Child Dev 58:741-749
Mélen M, Deliège I (in press) Extraction of cues or underlying harmonie structure: which guide recognition of familiar melodies? The European J of cognitive psychology
Mervis C.B. (1987) Child-basic object categories and early lexical development. In: Neisser U (ed) Concepts and conceptual development. Ecological and intellectual factors in categorization. Cambridge University Press, Cambridge pp 201-233
Neisser U (1987) From direct perception to conceptual structure. In: Neisser U (ed) Concepts and conceptual development. Ecological and intellectual factors in categorization. Cambridge University Press, Cambridge pp 11-24
Peeters T (1989) L'autisme: de la connaissance théorique à l'intervention éducative. In: Lelord G, Muh JP, Petit M, Sauvage D (eds): Autisme et troubles du développement global de l'enfant, recherches récentes et perspectives. Expansion Scientifique Française, Paris
Peretz I, Morais J (1989) Music and modularity. Contemp Music Rev 4:279-295.
Rosch EH (1975) Cognitive reference points. J Exp Psychol Gen) 104:192-233
Rosch EH (1976) Classifications d'objets du monde réel: origines et représentations dans la cognition. Bull Psychol (Special issue) "La Mémoire Sémantique" 242-250
Rosch EH (1978) Principles of categorization. In: Rosch E, Lloyd B (eds) Cognition and categorization. Erlbaum, Hillsdale pp28-49
Rosch EH, Mervis CB, Gray W, Johnson D, Boyes-Bream P (1976) Basic objects in natural categories. Cogn Psycho 3:382-439
Segui J (1992) Perception du langage et modularité. In Andler D (ed) Introduction aux sciences cognitives. Gallimard, Paris, pp 131-152
Segui J, Beauvillain C (1988) Modularité et automaticité dans le traitement du langage: l'exemple du lexique. In: Perruchet P (ed) Les automatismes cognitifs. Mardaga, Bruxelles, pp 13-25
Trehub SE, Thorpe LA (1989) Infants' perception of rhythm. Categorization of auditory sequences by temporal structure. Can J Psychol 43:217-229

The Zürich Performance Workstation Project - An Overview *

GUERINO MAZZOLA

The Performance Scheme: A Computerized Model of the Interpretation Process

If Susanne Langer is right in claiming that there is an isomorphism between musical structure and feelings, the concept of musical structure must include the work of interpretation. We think that Langer's hypothesis is very difficult to verify. To make our argument clear, I want to present the current work on the mathematical theory of interpretation by our research group at the Zürich Multi Media Lab. The principal result of this work is implemented in software called "performance workstation" on NEXTSTEP computers.

Any work of music is incorporated into a communication process which aims at conveying insights into the work's structure to a listener via instrumental performance. Our model is termed "interpretation scheme" (Fig. 1); it splits the interpretational communication process in two parts: First, the "neutral" score is analyzed by a specific choice of analytical instruments. We call this "structuring." The resulting "analysis" of a score is the effect of applying the given structuring to the score. The general frame wherein different structurings appear as special choices is called "structuring scheme;" the variables whose values determine different structurings are the "structuring variables." This makes clear that we intend to allow the user of our workstation to choose the analytical perspective for the given work! The structuring scheme will integrate analytical approaches such as Lerdahl and Jackendoff (1983), Maxwell (1992), Mazzola (1990), Rothstein (1989), Straus (1990).

Given a concrete analysis of the score, the interpreter has to set up an arsenal of interpretational and performance techniques which transform the analyzed score content into sounding music. In the second part of our model, this arsenal is called a "shaping." Its application to the given analysis is called "synthesis" and this one results in an effective "performance." The latter is realized via several possible physical devices, such as MIDI instruments or the IRCAM sound chart. The general frame in which different shapings may be positioned as special choices is called

*This work was supported by the Swiss National Science Foundation, grant no. 21-33651.92

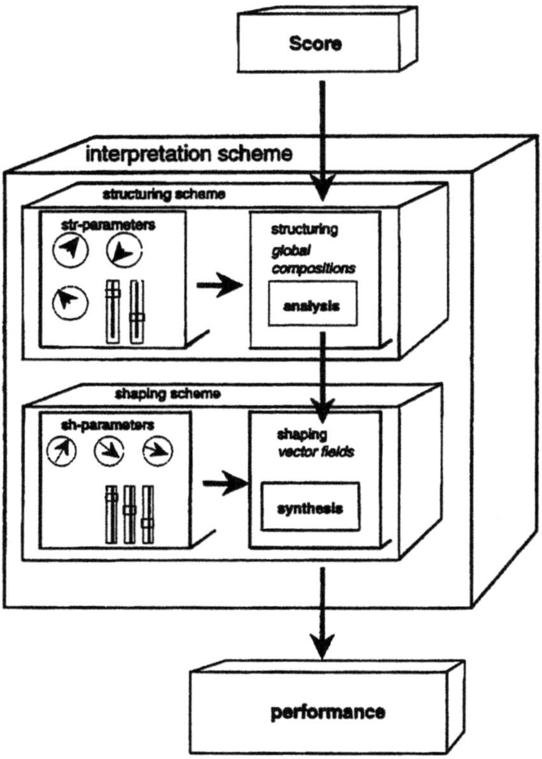

Fig. 1. The structure of the intepretation scheme

the "shaping scheme"; the variables whose values determine different shapings are the "shaping variables." Again, we should stress that these allow the user to construct his own perspective of shaping a given analysis! The shaping scheme integrates/generalizes performance theories and software concepts such as Friberg (1991), Friberg et al. (1991), Misch and Wille (1988), Abel and Reiss (1991), Mazzola (1989), Mazzola and Hofmann (1989), Mazzola and Muzzulini (1990), Sundberg et al. (1989), Sundin (1984).

Musical Expression via Vector Fields

Let me shortly explain how vector fields intervene in this context. Classically, the passage from symbolic musical time of the score data to the physical data describing performance is defined by the tempo curve as a function of symbolic onset time. A similar situation occurs when dealing with intonation: In this case, an intonation curve as a function of symbolic pitch is given. Looking simultaneously at tempo and intonation, a two-dimensional "performance vector field," the tempo-intonation field, is defined as a product of the one-dimensional factor fields tempo and intonation. In more sophisticated situations of expressive interpretation, this performance field is no longer split. However, it is shown that the values of the integral curve parameter determine the physical tone associated with a symbolic tone (a note of the score) and the performance field.

This situation can generally be applied to all basic parameters of onset, pitch, and loudness together with their "pianola" parameters of duration, glissando, and crescendo. The structure of the performance fields in general obeys four principles: Globalization, formation of hierarchies, factorization, and specialization (see Mazzola et al. 1993 for details). This complex structure of fields within the interpretation scheme already makes clear that a satisfactory mathematical model of interpretation and performance is something quite complicated, and that there is no reason to be afraid that mathematics will unduly simplify things. The decisive difference between the mathematical approach and the purely intuitive and metaphoric traditional approach in interpretation aesthetics is the precision, the operability, efficiency, and reliability of the former. However, classification of muscial structures including interpretation and performance is far from being controlled, and Langer's hypothesis, as far as it depends upon this classification, is presently too difficult to be verified.

References

Abel V, Reiss P (1991) MUTABOR II - Software manual. Mutabor Soft, Darmstadt

Friberg A (1991) Generative rules for music performance: a formal description of a rule system. Comput Music J 15 (2):56-71

Friberg A, Frydén L, Bodin LG, Sundberg J (1991) Performance rules for computer-controlled contemporary keyboard music. Comput Music J 15 (2):49-55

Lerdahl F, Jackendoff R (1983) A generative theory of tonal music. MIT Press, Cambridge/MA

Maxwell HJ (1992) An expert system for harmonizing analysis of tonal music. In: Balaban M, Ebcioglu K, Laske O (ed) Understanding music with AI: perspectives on music cognition. MIT Press, Cambridge/MA

Mazzola G (1989) PRESTO - software manual. SToA music, Zürich

Mazzola G (1990) Geometrie der Töne. Birkhäuser, Basel

Mazzola G, Hofmann GR (1989) Der Music Designer MD-Z71 - Hardware und Software für die Mathematische Musiktheorie. In: Petsche H (ed) Musik - Gehirn - Spiel. Birkhäuser, Basel (4th Herbert-von-Karajan-Symposion)

Mazzola G, Muzzulini D (1990) Tempo- und Stimmungsfelder: Perspektiven künftiger Musikcomputer.In: Hesse HP (ed) Mikrotöne III. Helbling, Innsbruck

Mazzola G (1993) Geometry and logic of musical performance. Swiss National Science Foundation Report, grant no. 21-3365.92. University of Zürich

Misch C, Wille R (1988) Stimmungslogiken auf MUTABOR: Eine Programmiersprache. In: Herf FR (ed) Mikrotöne II. Helbling, Innsbruck

Rothstein W (1989) Phrase rhythm in tonal music. Schirmer, New York

Straus JN (1990) Introduction to post-tonal theory. Prentice Hall, Englewood Cliffs

Sundberg J, Friberg A, Frydén L (1989) Rules for automated performance of ensemble music. Contemp Music Rev 3:89-109

Sundin NG (1984) Musical interpretation in performance. Mirage, Stockholm

II. Music Physiology and Pathophysiology

Music Perception in the Auditory Hierarchy

ERNST TERHARDT

Introduction

For hundreds or even thousands of years man has been fascinated by the systematic organization that is so readily apparent in the sounds of music. No other acoustic medium of communication exists which is so rigorously and precisely organized in both time and pitch. In the time - domain, we observe that patterns of successive acoustic events are appreciated by the auditory system as holistic entities, and that such entities often are subdivided into - or composed of - subentities. In other words, in music, temporal patterns are to a most remarkable extent hierarchically organized. As regards the low-high domain, we observe that musical pitch is defined by categories, instead of forming a continuum. As a consequence, musical instruments have evolved which are only capable of producing pitches that are within the pitch categories of our tone scale, yet are powerful tools for both the composer's and the performer's musical expression. Moreover, we observe that pitch intervals exist which have certain singular auditory qualities, such as the octave and the fifth, and we notice that those particular intervals are, in a sense, dependent on ratios of small integers such as 1:2 for the octave, 2:3 for the fifth, and so on. Eventually, we observe that many simultaneous or successive tones have what may be termed "harmonic implications," i.e., "harmony" in the functional rather than aesthetic sense. As was first conceptualized by Rameau (1750), the harmonic implications of musical sound can be depicted by corresponding bass notes (*Basse fondamentale*). These phenomena are hierarchically organized as well, as will be pointed out in more detail below.

As compared to ordinary sounds of daily life - including human speech - the systematic organization of musical sound is so pronounced and rigorous that one may be inclined to believe that music is an artificial and somewhat arbitrary invention of man. However, while there cannot be any doubt that music indeed is an invention of man which for thousands of years has been - and still is - in continuous development, there are good reasons for doubting that it is entirely artificial and arbitrary. Instead, the picture emerges that music not only obeys the basic and general principles of sensory acquisition of information but that it even provides an archetype of those principles, i.e., exhibiting them most purely. The present contribution attempts to illustrate and support this assertion.

The Hierarchy of Temporal Patterns

By "temporal pattern" we mean a sequence of events. An event, in turn is any segment of a sound signal that is heard as occurring at a specific instant in time. As a dimension of auditory perception, time has two somewhat different aspects. First, it naturally is the dimension in which auditory events follow one another. Second, it plays the role of a Gestalt dimension. This notion is essential for understanding the phenomenon of rhythm, i.e., of the hierarchical grouping of events.

In music, "events" indeed exist on quite different hierarchical levels. On a high level, an evening on which three symphonies are performed can be regarded as an event, not only in a metaphoric sense. Each of the symphonies in turn then is a subevent; each of its movements is another subevent. Stepping down through the hierarchy we find: a particular movement's sections; a section's musical phrases; and the tones of the phrase. Even here the chain of hierarchical levels does not end, as a musical tone ordinarily has temporal structure es well, i.e., it is composed of audible subevents.

From psychoacoustic data on the perception of roughness, auditory flutter, and the like, one can conclude that there exists a type of elementary auditory event such that it cannot be subdivided into audible subevents. The temporal extension of those elementary events turns out to be on the order of a few milliseconds.

The Auditory Hierarchy of Processing Temporal Patterns

As regards auditory perception, we can indeed appreciate all or several of those hierarchical aspects of temporal patterns. We can hear the temporal structure that exists within a single tone, such as vibrato, yet we can also comprehend the rhythmical pattern established by a number of successive tones. We also more or less readily put that pattern into the context of the preceding and following ones.

Moreover, we can aurally maintain control even of long uniform sequences of events, i.e., patterns which do not include any clues to subpatterns. Patterns of equidistant events are on some level of the auditory hierarchy "automatically" grouped into subgroups of, for example, two, three, or four events. The practical advantage of this achievement both for ordinary acoustic communication and for the performance and appreciation of music is that one can maintain perceptual control even of long uniform patterns without counting the individual events.

It is impossible to understand these perceptual processes in the frame of a model of auditory perception of the "one-input-one-output" type. Rather, while there indeed is just one input per ear, there are several outputs, i.e., corresponding to different hierarchical levels of the auditory system.

Notes on Relationships Between Music and Speech

With respect to temporal patterns of events, it is elucidating to compare music with speech. In fluent speech, the most prominent quasi-elementary events are the syllables. Listeners who are asked to reproduce the rhythm of natural fluent speech by tapping, produce a pattern that matches that of syllables (Köhlmann 1984a). In normal speech, the number of syllables per second is on the average three to five. When one divides the total number of (successive) tones of any piece of music by the time typically elapsed in its performance, one typically gets figures in the range of 1.7 to 5.5 tones per second. Moreover, psychoacoustic tapping experiments with music of various types yields a distribution of the time intervals between successive auditory events which is exactly in the same range (Köhlmann 1984b). So, as elements of temporal patterns, musical tones can indeed be concluded to correspond to syllables of speech.

A corresponding analogy exists for "utterances" (short sentences). In many types of music one can find "utterances" which, on the average, correspond both in their duration and the number of tones which they include to utterances of speech (where the above analogy between tones and syllables is employed).

Evidently music differs from speech in its more rigorous and regular temporal organization of events. On first sight this feature of music may indeed appear as arbitrary, artificial, and of no practical value in acoustic communication. However, closer inspection reveals that this is not so. To elucidate this assertion, it is helpful to consider the relationship between two basic types of human vocalization: Natural speech, and song.

Speech Versus Song

As was emphasized already by Helmholtz (1863), singing provides a most significant psychophysical link between speech and music. A closer inspection of what, physically, is the difference between "normal" speech and song may yield valuable insights. The most apparent differences are:

- Elongation of vowels and phonation of each syllable with an essentially constant, pronounced voice pitch such that syllables correlate with pitches.
- Selection of voice pitches that are in a simple musical relationship such as a minor third.
- Use of more elaborate, i.e., artificial, rhythms.

It is self-evident that shouting (i.e., "primitive singing") is employed for improving communication by voice, for example, in a noisy environment. Most interestingly, the psychophysical features which provide for that improvement are to a considerable extent identical with what is typical for music: Elongation of vowels (so that they dominate the syllable and become "tones"); use of distinctly and fairly systematically different (i.e., categorized) pitches; and creation of well-organized, somewhat artifical temporal patterns. This makes apparent that it is exatly those seemingly artificial and arbitrary features of music that have considerable "survival value" because they are efficient in improving the "signal-to-noise ratio" in communication by voice.

Hierarchy in the Pitch Domain

The auditory low - high continuum is essentially the dimension of pitch; therefore it is briefly termed the pitch domain. It is ordinarily very complex, because for the vast majority of sounds there exist at every instant of time a considerable number of pitches at once. Moreover, multiple pitches sound simultaneously on different levels of the auditory hierarchy. Pitch, in general, is far from being a simple, elementary auditory attribute. The complexity which in fact is included in pitch is suggested by the variety of concepts which exist about it.

Traditionally, one tends to conceive the relationship between stimulus parameters and perceived pitch as a straightforward input - output relationship (in an extremely simplicistic approach, as a kind of synonymity of pitch and frequency). This approach has turned out to be one of the most serious obstacles to understanding pitch perception. From many everyday observations it can be made apparent that the pitch of perceived "tonal objects" is multiple, i.e., both "horizontally" (within one hierarchical level) and "vertically" (across levels).

As an example, consider the sound of a note played on a pipe organ in a mixture register. Pressing the pertinent key will cause several pipes to sound simultaneously with related oscillation frequencies that are nearly in ratios of small integers. On a moderately high level of the auditory hierarchy - to which attention normally is preset - one indeed hears a single "note." If one switches attention to the next lower level, one can notice the individual pitches of the pertinent pipes. So what on the higher level is heard as one single object is on the lower noticed as a set of subobjects. On yet another level below, one may even hear some of the sinusoidal part tones of which each of the pipe sounds is (theoretically) composed - i.e., sub-subobjects of the musical note. The same hierarchical relationship exists between virtual pitch and spectral pitch.

Hierarchical Perception of Pitch Implies Musical Tonality

The notion of the hierarchical organization of pitch provides us with a key to understanding no less than the origin of tonality, consonance, and harmony. For example, Rameau's *Basse fondamentale* turns out to be a "real," i.e., psychophysically based, phenomenon, namely, virtual pitch. In other words, those features of tonal music stringently emerge from the principles of pitch perception. This includes an explanation of why and how small ratios of integer numbers come into play whenever one is concerned with musical sound.

As these explanations have been given in detail on several earlier occasions (e.g., Terhardt 1984), we can focus our attention to the question, which is the "survival value" of auditory formation of pitch?

The Role of Pitch in Acoustic Communication

The important role that pitch plays in music is an immediate consequence of its role in ordinary acoustic communication, and this in fact is a key role. To understand this, it is only necessary to take account of both what is the essential purpose of the ear, and what the physical conditions for fulfilling that purpose are.

The purpose, which is defined by success in natural selection, of course is efficient acquisition of acoustic information. More precisely, acquisition of information about what kind of objects exist in the environment, what they are doing, what they possibly intend to do, and how they possibly can be used. Therefore, from the physical parameters of the auditory stimulus (e.g., amplitude, phase, and frequency) the most useful - and therefore important - are the most robust carriers of information coming from sound sources in a distance. It is not difficult to show that in a frequency-domain signal description it is the frequencies (as opposed to phases and amplitudes) which are most robust (see Huggins 1952). In particular, spectral frequencies are involved such as:

- The frequencies of decaying eigen-oscillations of sound objects, for instance, those of a bell, and of a string;
- The frequencies of harmonic part tones of steady periodic sound signals such as those the human voice

From the combination of physical facts, namely, (1) that frequencies carry a maximum of information about the physical state and activity of sound sources, and (2) that they are highly robust in physical transmission of sound, there emerges a pronounced selective pressure on the evolution of hearing organs that can perceive pitch. The human auditory system is a product of that selective pressure. What we can hear as spectral pitches of sound in fact are an accurate representation of spectral frequencies that through their appearance and disappearance in time and their temporal variation provide reliable and robust information about the sound sources of the environment. The spectral pitches in turn provide a most essential basis of the formation of higher-level auditory attributes such as virtual pitches - not to forget timbre. Once attention has been drawn to the importance of pitch for "plain" acoustic communication, it is not too difficult to become aware of the fact that practically any type of sound includes pitches. It is only that we often do not become aware of their existence in sounds that, superficially, appear to be more or less noise - like - not to mention that we actually draw the bulk of information from those pitches. The seeming paradox that we are not usually aware of pitches which are so important carriers of information is resolved by the notion of hierarchical perception: Lower levels of the auditory hierarchy are autonomous enough to take advantage of cues that we do not consciously notice, although we can notice them if we focus our attention carefully. It is in this sense that "subliminal" cues of sound are of highest functional significance. Particularly with respect to music the latter notion is familiar to most experienced listeners.

Generalizations and Conclusions

In the foregoing sections we have illustrated that the rigorous and systematic organization of music both in the time and the low - high domain is neither arbitrary nor purely artificial, but rather is archetypical for the essential principles of the auditory acquisition of information. Those principles, in turn, can be put into the context of sensory information acquisition in general. Common to all systems of information acquisition and -processing are, briefly, the principles of discretization, hierarchical organization, and knowledge-based decision on all levels of the hierarchy.

Knowledge-based decision-making on all hierarchical levels implies that each of the hierarchical layers of processing is to a considerable extent autonomous. Discretization in the time domain becomes manifest in the notion that an essentially continuous signal (such as speech) is perceived as a sequence of events. In the pitch domain, discretization is manifest in the mere existence of discrete pitches. In vision, an analogous case of discretization is the formation of contours. In fact, for visual acquisition of information contours play exactly the same role as was outlined above for spectral pitches: They convey robust information about external objects. In this sense one may regard pitch as "auditory contour." Likewise, temporal events can be termed "temporal contours." And several types of contour ordinarily appear on any level of the sensory hierarchy in such a way that higher-level contours are in a more or less complex way dependent on lower-level contours (see Terhardt 1989; 1991).

In these general terms, music appears as a "play with contours" which obeys and reflects the universal principles of sensory acquisition of information.

References

Helmholtz H (1863) Die Lehre von den Tonempfindungen als physiologische Grundlage für die Theorie der Musik. Vieweg, Braunschweig

Huggins WH (1952) A phase principle for complex-frequency analysis and its implications in auditory theory. J Acoust Soc Am 24:582-589

Köhlmann M (1984a) Bestimmung der Silbenstruktur von fließender Sprache mit Hilfe der Rhythmuswahrnehmung. Acustica 56:120-125

Köhlmann M (1984b) Rhythmische Segmentierung von Sprach- und Musiksignalen und ihre Nachbildung mit einem Funktionsschema. Acustica 56:193-204

Rameau JP (1750) Traite de l'harmonie. Ballard, Paris

Terhardt E (1984) The concept of musical consonance: a link between music and psychoacoustics. Music Percept 1:276-295

Terhardt E (1989) Warum hören wir Sinustöne? Naturwissenschaften 76:496-504
Terhardt E (1991) Music perception and sensory information acquisition: relationships and low-level analogies. Music Percept 8:217-239

Neuronal Response in Limbic and Neocortical Structures During Perception of Consonances and Dissonances

Guerino Mazzola

Introduction

The aim of the project at the Neurology Department of the University Hospital at Zürich which H.G. Wieser, the author, and their collaborators conducted during the years 1984-88 was to test aesthetic principles of classical counterpoint by means of depth EEG responses to musical stimuli. In particular, it was planned to test the relative results of the EEG to different musical inputs (consonances vs. dissonances, to fix the ideas) and not the relation of responses to musical stimuli versus nonmusical stimuli. The latter problem is being investigated with much success by H. Petsche and his collaborators (see Petsche et al. 1989 and his contribution in this volume). It is important to stress that from our results, we do not draw any kind of conclusions concerning a possible genetic nature of musical understanding or a possible universal validity of classical European interval categories. Our investigations show that in some defined regions of the brain of some European humans, certain significant reactions take place - nothing more and nothing less. There is no reason to generalize whatsoever, but there are enough reasons to try to repeat these investigations in other research sites with a comparable infrastructure.

For a complete report of our results, we refer to Wieser and Mazzola (1986, 1987), Mazzola et al. (1989a,b). In this short report, we shall restrict ourselves to the two subtests concerning (a) isolated successive intervals and (b) the autocomplementarity function between simultaneous consonances and dissonances. Our results confirm our hypothesis on (1) a significant differentiation of EEG responses to consonant vs. dissonant intervals in limbic and auditory brain areas and (2) a pronounced sensitivity of these areas in EEG responses to the fundamental autocomplementarity relation between consonances and dissonances. In particular, the quantitative measurement of these responses by use of the "spectral participation vector" has confirmed our belief that this vector may carry some of the semantic charge of EEG signals.

Aesthetic Principles of Counterpoint

In our project, we dealt with two aesthetic principles of classical counterpoint as it is described in Fux (1725) and following the final system of Giovanni da Palestrina. The first principle concerns the dichotomy of the set of intervals into a consonant and a dissonant half. The second is the variation principle, a fundament for formation of contrapuntal rules.

It is crucial to concentrate on a specific layer of reality when discussing a musicological question since, in each layer of reality, a different answer may be given to one and the same question. Take, as an example, the concept of consonance and dissonance. On the physi(ologi)cal level, as investigated by von Helmholtz (1863/1968) in his beating theory, simple quotients of frequencies like 2/1 (octave), 3/2 (fifth) or 4/3 (fourth) yield more "consonant" intervals than complicated fractions such as 45/32 (tritone). On the psychological level as investigated by Plomp and Levelt (1965), this contrast no longer holds. On the symbolic/mathematical level as inquired by Euler (1739/1926) through his number theoretic gradus suavitatis function, we get yet another measure of more or less agreeable intervals. However, it is crucial to recognize that the mutual contradiction of these approaches is only apparent since they are tied to distinct layers of reality.

Note that none of these concepts furnishes a dichotomic division of the set of intervals; they always are more or less agreeable. In our approach, we prefer sticking to the clear-cut symbolic/musicological point of view of composers and music theorists. It is the only approach in which a sharp dichotomy of intervals is defined. And we do prefer this one because it is not a matter of telling composers and music theorists what consonances have to be, but rather we want to understand why they came to finally define the consonanance - dissonance dichotomy as it is. Within an octave, consonances are the prime, the minor and major third, the fifth and the minor and major sixth. Minor and major seconds, the fourth, tritone, and minor and major seventh are dissonant. This is the core of counterpoint theory as described by Sachs (1974) (first species: two voices, note against note).

The Variation Principle

Originally, the cantus firmus melody from old Gregorian choral was enriched by adding a second parallel voice, the descant, a fifth or fourth above (sweeping) or below (hanging) the cantus firmus. This boring construction developed into Palestrina´s very sophisticated system of rules made in such a way that a maximum variation of intervals could be achieved. The most important rule in this system is the rule of forbidden parallels of fifths, an evident prescription preventing the composer from building boring constructions. This variational principle fits perfectly with Sachs´ insight that "punctus contra punctum" not only means to set a descant note against a cantus firmus note, but also to set the successor of a given interval as being a counterpoint to its predecessor: "contra" meaning "iuxta," a contraposition in time.

Of course, this variation principle is easier to state in general aesthetic terms than to realize in a coherent system of concrete rules, all the more so since the movement from interval to interval has to take place within the subset of consonant intervals, and the given contrast of consonance vs. dissonance cannot be used in a straightforward way.

A Counterpoint Model in Mathematical Music Theory

Using mathematical models of counterpoint, we have tried to establish a general formal principle of variation which is appropriate to deduce the rules of counterpoint, at least those of the core theory according to Sachs.

The fundament of this model is the fact that the classical dichotomy is glued together by a unique inner symmetry of the interval set, the so-called autocomplementarity function (AC function). This operates as follows: If we fix the twelve intervals modulo octave to be the integers modulo 12 (prime=0, minor second=1,....major seventh=11), then we can represent the interval set as being twelve points on the torus of thirds (see Fig. 1).

They are distributed on three horizontal circles, equidistant to each other by a major third, or on four meridians, equidistant to each other by a minor third. The AC function is a linear expression given by $y=5x+2$; it transforms the consonant prime $x=0$ to the dissonant major second $y=2$, and so on; the complete list of AC-related consonances x and dissonances y is $(x,y)=(0,2)$, $(3,5)$, $(4,10)$, $(7,1)$, $(8,6)$, $(9,11)$. Conversely, the same functional expression associates a dissonance x with a consonance y, and this relation is just the inverse of the former one, i.e., $(x,y)=(2,0)$, $(5,3)$ etc. There is no other linear expression which transforms the set K of consonances into the set D of dissonances!

Of course, there are many other partitions of the total set of intervals into two six element subsets. But the K-D partition is distinguished for several mathematical

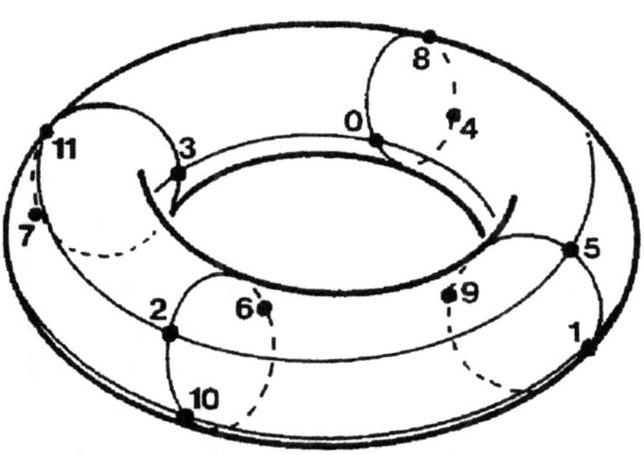

Fig. 1. On the torus of thirds, the consonance - dissonance dichotomy is shown. We recognize the characteristic total separation of consonances from dissonances into two connected regions. *Numbers* refer to the intervals as explained in text

reasons. One of them is quite evident from Fig. 1: On the torus of thirds, K is separated from D in an optimal way; for more details, see Mazzola (1990).

Deformation Forces by Local Symmetries

The idea of our counterpoint model was to deform the given K-D dichotomy by use of special symmetries. Why? From counterpoint theory as described by Apfel (1982), Fux (1725), Sachs (1974) and other theorists, it became clear that the variation principle means that you in some sense try to move from consonances to dissonances within the set K of consonances - a contradictory requirement, of course, since the first species of counterpoint absolutely demands to move within the subset of consonants.

Here is the point of our construction: Deformations displace the "separation line" between consonance and dissonance. This means that we now have separated certain consonances from the remaining consonances by use of this deformed separation line. The principle of variaton will then apply in the sense that a transition from a given consonance to another is subjected to the condition of crossing the deformed separation line! This idea enables us to establish complete lists of allowed or forbidden transitions from interval to interval (see Mazzola et al. 1989a and Mazzola 1990). These transitions coincide very well with those indicated by Fux (1725), as has been verified by my collaborator, the musicologist D. Muzzulini (see Mazzola 1990).

The EEG Tests

Our test concerned different contexts of consonances and dissonances as well as the test of the AC function. We used EEG from the scalp (Hess system), stereotactic depth EEG following Wieser (1983), and multipolar foramen ovale recordings (see Wieser and Moser 1988). The tests were applied to the rare cases of patients suffering from medically intractable complex partial epilepsy seizures of suspected mediobasal temporal lobe origin and underwent presurgical evaluation with a view towards surgical epilepsy therapy. None of the total of 13 patients considered the voluntary 30 min music test through monophonic earphones as being disagreeable.

There are several reasons why, notwithstanding the particular state of epileptics, the tests remain comparable to tests with normal humans which cannot be conducted for evident reasons. First, our tests were performed during interictal periods. Second, localization of the focus gives a good estimation of its possible influence. Third, epileptiform potentials are easily distinguished from others by the expert.

For each patient, we recorded 700 time windows for fast Fourier transform (FFT) spectral analysis with 256 samples per second, each window for different EEG channels, different power windows (delta 0-4 Hz, theta 4-8 Hz, alpha 8-14 Hz, beta 14-40 Hz) and repetitions, totally 11 000 raw spectral data per patient. Unfortunately, the project could not be completed for extrascientific reasons and hence, only two

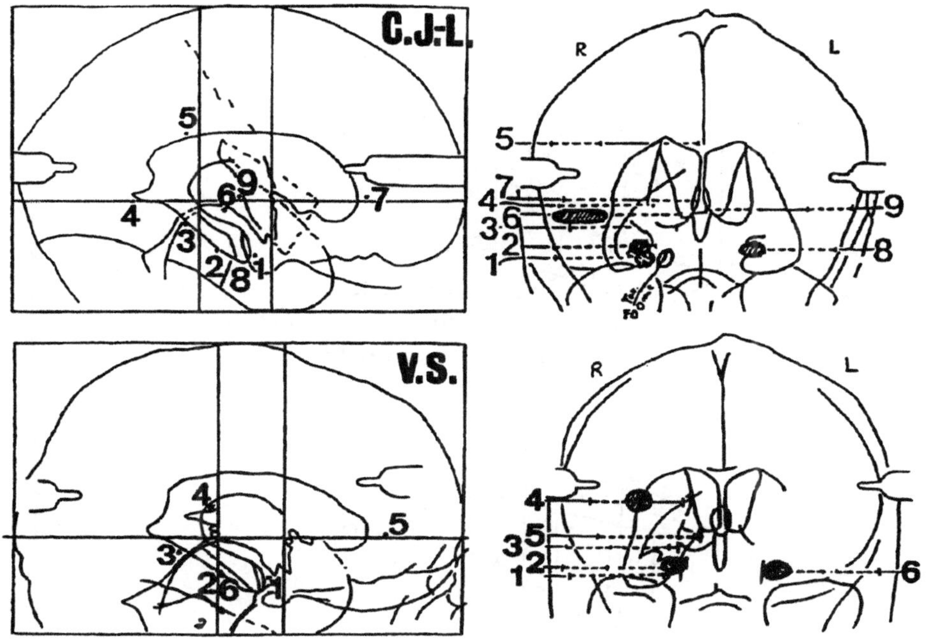

Fig. 2. Implantation scheme for the two patients C. J.-L. and V.S. following X-rays showing the topographic position of depth electrodes. The three homologous regions are indicated by *lines*. For C. J.-L., RCA is electrode 2/1-3, RH is 6/5-6 and LCA is 8/1-3. For V.S., 4 is electrode 2/1-2, 10 is 4/5-6 and 14 is 6/1-2. (Numbers after the slashes indicate precise positions on the electrodes, where 1 = deepest position and 10 = position near surface)

patients have been thoroughly evaluated. These patients were C. J.-L., a 35-year-old academic, and V.S., a 31-year-old artisan. Both are male Europeans and prefer standard classical and standard light and folklore music. Figure 2 shows the positions of bipolar depth EEG recordings which we are going to discuss. Notice that C. J.-L.'s recordings RCA, RH, LCA are homologous to V.S.'s recordings 4, 10, 14. Recording RCA lies within the right hippocampus, recording RH is positioned within Heschl's gyrus (auditory cortex), and recording LCA lies within the left hippocampus.

Nevertheless, we have been able to observe a great deal of visual evidence for EEG responses to music stimuli in neo- and archicortical regions of all patients; for details, see Mazzola et al. (1989a) as well as Wieser and Mazzola (1986).

Analysis by Spectral Participation Vectors

For our tests, we used four sounds: piano, sine wave, cello (without vibrato) and "test," a clear, organ-like sound, all synthesized from Yamaha TX7 synthesizer and CX5M voicing program in order to avoid possible emotional artifacts associated

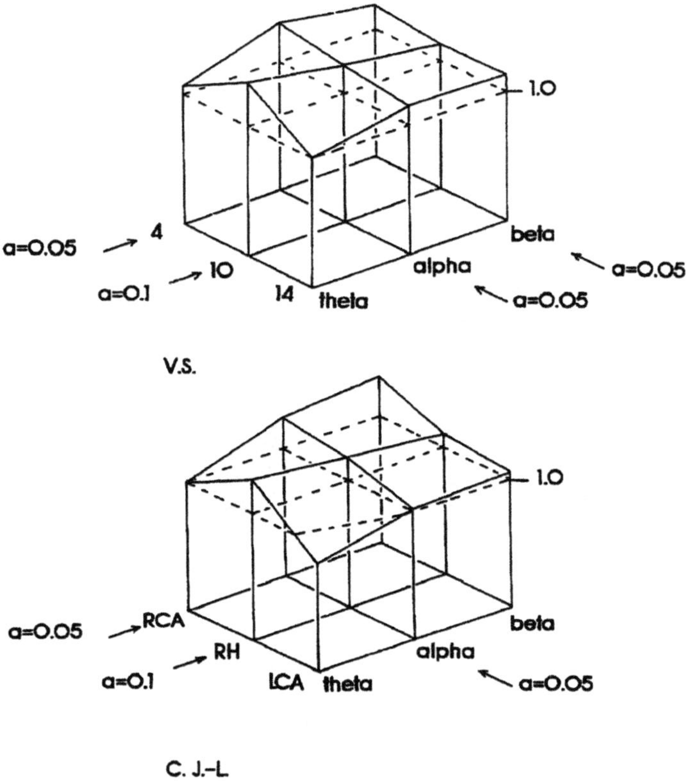

Fig. 3. Graphic representation of the quotients *Qth(K,D)*, *Qa(K,D)* and *Qb(K,D)* for all locations and patients C. J.-L. and V.S.. These *numbers* show that for con-sonances, participation lowers more when the second tone appears than for dissonances: observe the cases where the respective number is greater than 1

with natural sounds from sociocultural premises. The music program was written on a precursor of the commercial composition software presto, now available for Atari computers (see Mazzola 1989).

The spectral analysis was executed on a CDC Cyber. We made use of the spectral participation vector $S(E)=(P(E), P(E)/Pd(E), P(E)/Pth(E), P(E)/Pa(E), P(E)/Pb(E))$ of an event E and its associated participation value $v(E)=P(E)/Pth(E)+P(E)/Pa(E)+P(E)/Pb(E)$. Here, $P(E)$ is the total spectral power, $Pd(E)$ is the delta power, $Pth(E)$ is the theta power, $Pa(E)$ the alpha power and $Pb(E)$ the beta power of E, see Mazzola et al. (1989) for details.

From the results obtained thus far we conclude that this representation is well suited to give an adequate picture of possible semantical charge of EEG signals. It is a measure motivated by, among others, the well-known vigilance-related alpha participation *Sa(E)=P(E)/Pa(E)* observed by Berger. We also use the delta participation *Sd(E)=P(E)/Pd(E)*, the theta participation *Sth(E)=P(E)/Pth(E)*, and the beta participation *Sb(E)=P(E)/Pb(E)*.

Isolated Successive Intervals

We first focus our attention on a subtest concerning musically isolated successive intervals, i.e., the two tones of an interval are played one after the other without interruption. All the intervals were played in three orders: (1) All consonances, ordered according to their size; then all dissonances, also ordered according to their size. (2) All consonances, ordered according to complementary size (if possible); then all dissonances, ordered according to complementary size (if possible). (3) A mixed succesion of all intervals according to a particular dodecaphonic all-interval series.

Having fixed a frequency band, theta, say, and an interval *E*, we take the quotient *Qth(E1,E2)=Sth(E1)/Sth(E2)* of the theta participations of the first tone *E1* and the second tone *E2*. If *Qth(E1,E2)*>1 or <1, theta participation lowers or increases from the first to the second tone. In order to compare these ratios for consonances and dissonances, we take the quotient *Qth(K,D)=Qth(K)/Qth(D)*, where *Qth(K)* or *Qth(D)* is the mean value of all Qth(E1,E2) for consonances or dissonances. This construction is repeated for all recording positions and all frequency bands, inclucing theta, alpha and beta; delta is omitted since it may be affected by noise. This test was performed four times with patient C. J.-L. and six times with V. S. A one-sided Wilcoxon test shows significantly higher quotients for consonances compared to dissonances, i.e. *Q??(K,D)* is significantly larger than 1 for many recording positions and frequency bands; see Fig. 3, and observe the similarity of distribution of these quotients for our patients with respect to homologous recordings and frequency bands. This means that for consonances, participation lowers more when the second tone appears than for dissonances.

Autocomplementarity Function

To test the response to the AC function, we confronted each consonant interval *X* with all the dissonant intervals *Y* and we looked for particular responses in cases, where *Y* was the interval which should correspond to *X* according to the formula $Y=2X+5$ of the AC function. Here, we looked at simultaneous intervals. For each given consonance *X*, we played a sequence of six confrontations, i.e. immediate successions *(X,Y1), (X,Y2),... (X,Y6)* of *X* with each of the six dissonances *Y1,Y2,...,Y6*. The duration of each interval was 0.68 s. We looked for the least participation values among the dissonant intervals *Y1,Y2,...,Y6*, when confronted with a fixed consonance *X*. We then compared the effective hits to the a priori chance to hit the correct dissonance. This method was applied to every recording position and to the

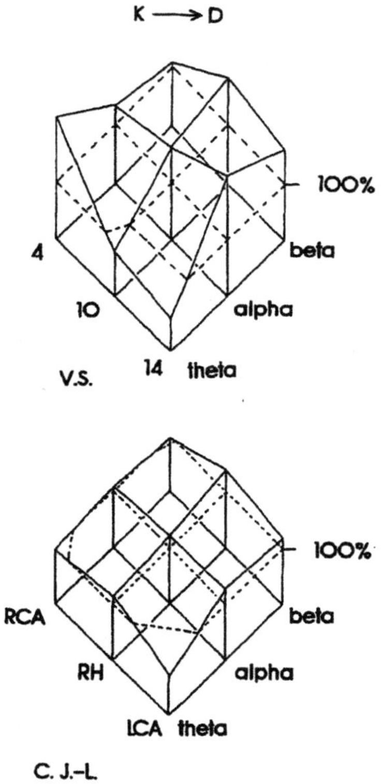

Fig. 4. Graphic representation of hitting frequencies for the auto-complementarity (AC) function subtest for patients C. J.-L. and V.S., three homologous locations as well as theta, alpha and beta frequency bands. The frequent and pronounced values above 100% show that for both patients there was a strongly affirmative EEG response to the test of the correct values for the AC function. In addition the topographic/spectral distribution of values for alpha and beta bands is comparable for these patients

three above frequency bands. Figure 4 shows the numbers compared to 100%, the measure for a priori chance. Due to the small number of samples for this subtest, we did not apply any statistical test here. However, as Fig. 4 shows, the results are remarkable and similar for both patients, and we conclude that this pilot investigation strongly supports the presence of the AC function as a foundation of contrapuntal processes.

References

Apfel E (1982) Diskant und Kontrapunkt in der Musiktheorie des 12. bis 15. Jahrhunderts. Heinrichshofen, Wilhelmshaven

Euler L (1926) Tentamen novae musicae. In: Bernoulli E et al. (eds) Opera omnia, ser III, vol 1. Teubner, Leipzig (Original 1739)

Fux JJ (1725) Gradus ad Parnassum. German and commented by L. Mitzler, Leipzig 1742

Mazzola G (1989) PRESTO - software manual. SToA music, Zürich

Mazzola G (1990) Geometrie der Töne. Birkhäuser, Basel

Mazzola G, Wieser HG, Brunner V, Muzzulini D (1989a) A symmetry-oriented mathematical model of classical counterpoint and related neurophysiological investigations by depth-EEG. In: Hargittai G (ed) Symmetry II, CAMWA. Pergamon, New York

Mazzola G, Graber-Brunner V, Wieser HG (1989b) Hirnelektrische Vorgänge im limbischen System bei konsonanten und dissonanten Klängen. In: Petsche H (ed) Musik - Gehirn - Spiel. (4th Herbert-von-Karajan-Symposion), Birkhäuser, Basel
Muzzulini D (1990) Konsonanz und Dissonanz in Musiktheorie und Psychoakustik. Musikwissenschaftliches Seminar der Universität, Zürich
Petsche H, Lindner K, Rappelsberger P, Gruber G (1989) Die Bedeutung des EEG für die Musikpsychologie. In: Petsche H (ed) Musik - Gehirn - Spiel, (4th Herbert-von-Karajan-Symposion), Birkhäuser, Basel
Plomp R, Levelt W (1965) Tonal consonance and critical bandwidth. J Acoust Soc Am 38:548
Sachs KJ (1974) Der Contrapunctus im 14. und 15. Jahrhundert. Steiner, Wiesbaden
Squire LR, Butters N (eds) (1984) Neuropsychology of memory. Guilford, New York
von Helmholtz H (1968) Die Lehre von den Tonempfindungen als physiologische Grundlage der Musik. (Orginal 1863). Olms, Hildesheim
Wieser HG (1983) Electroclinical features of the psychomotor seizure. Fischer, Stuttgart and Butterworth, London
Wieser HG, Mazzola G (1986) Musical consonances and dissonances: are they distinguished independently by the right and left hippocampi? Neuropsychologia 24 (6):805-812
Wieser HG, Mazzola G (1987) EEG responses to music in limbic and auditory cortices. In: Engel J Jr, Ojemann GA, Lüders HO, Williamson PD (eds) Fundamental mechanisms of human function. Raven, New York
Wieser HG, Moser S (1988) Improved multipolar foramen ovale electrode monitoring. J Epilepsy 1:13-22

Neural Nets and Music Cognition

Jamshed J. Bharucha

Introduction

Psychological research in auditory perception has been polarized into psychoacoustics and cognitive psychology. Yet segregating "low level" and "high level" processing prevents us from exploring the existence of common principles. In this paper I will briefly sketch how similar principles of learning and processing may underlie auditory phenomena at many levels. This approach would compel us to question the easy invocation of mental representations when psychoacoustic phenomena may suffice. It would also compel a renewed examination of psychoacoustic phenomena that are context sensitive and thought to be processed centrally rather than peripherally.

The proposed common principle encompassing many low and high level auditory processes is neural self-organization. Self-organizing or unsupervised neural nets have been studied by Grossberg (1970, 1972, 1976), von der Malsberg (1973); and Fukushima (1975), and more recently by Kohonen (1984) and Rumelhart and Zipser (1985). They are capable of pattern recognition and pattern completion following passive perceptual learning. In the study of music cognition, networks of this kind have been employed elegantly by Gjerdingen (1989, 1990) to explore the learning and recognition of sequential patterns that characterize musical styles. These models are in the spirit of early neural models that map auditory features onto more abstract musical categories (Deutsch 1969; Pitts and McCulloch 1947).

There is considerable debate about the theoretical status of neural net models, so a word is in order about the role they play in our research. Recent advances in our understanding of neural nets have pertained to their ability to learn. The logic of the theoretical perspective advanced here is thus as follows. Given fundamental constraints on neural architecture and general purpose neural learning mechanisms, and given passive extended exposure to an auditory environment that is structurally constrained - physically as well as culturally: (1) what aspects of this structure can plausibly be learned, and (2) what phenomena emerge as an automatic consequence of how these learned codes influence subsequent perception? If phenomena automatically emerge from the passive learning of environmental structure, then, on of grounds parsimony, the learning hypothesis is to be preferred to the postulation of innate or ad hoc mechanisms.

The models discussed here are primarily models of learning and computational process based on general neurophysiological assumptions and only secondarily models of actual neural connectivity. That is not to say that such models do not also serve as models of specific neural connectivity. In particular, they generate hypotheses about the kinds of neural feature detectors to look for when recording from the brain, and they also provide hypotheses for the kinds of dissociations one might look for in studies of neurological patients.

Most importantly, neural nets have pattern completion properties that lead to Gestalt phenomena. After becoming familiar with patterns, a network will complete or fill out a partial or degraded pattern. This approach is thus consistent with many of Terhardt's (1974) insights.

I shall sketch how self-organization can be employed to understand how the extraction of chord and key information is learned and how the extraction of pitch is learned. The first is a "high level" - traditionally cognitive - process and the second is a "low level" - traditionally psychoacoustic - process.

Neural Self-Organization

Neural self-organization is a learning procedure ideally suited to passive perceptual learning. In the context of the present work, it can be sketched as follows. One assumes a layer of neural units (input units) that are feature detectors. In the network that learns to extract chord and key information, these are octave-equivalent pitch units. In the network that learns to extract pitch, these are frequency-tuned cochlear neurons.

Each input unit is connected to each unit in a second layer of units (output units) that initially do not have any particular tuning or feature-detecting properties. Self-organization consists of changing the strengths of the synaptic connections (weights) from the input to the output so that the output units become tuned to (feature detectors for) entire patterns of activation in the input.

Each output unit has a vector of weights feeding into it (one weight for every input unit). Each stimulus pattern is a vector of activations across the input units. The weight vectors of the output units and the activation vectors of the stimuli can be represented in the same vector space with appropriate normalization, since both have the same number of elements (i.e., the number of input units). The output unit that is activated most strongly by a stimulus is the one whose weight vector is closest in angle to the stimulus vector. This winning output unit gets to learn. Learning is essentially Hebbian (Hebb 1949): weights to the winning output unit are strengthened if they connect with strongly activated input units and are weakened if they connect with weakly activated input units. This increases the likelihood that the same output unit will win in response to this stimulus and similar stimuli and reduces the likelihood that it will win in response to dissimilar stimuli. This output unit thus becomes tuned to this stimulus.

After learning a set of stimuli, the network becomes a pattern recognition device that will recognize these stimuli as well as stimuli similar to them. It will also endow degraded or novel stimuli with the properties of the learned stimuli to which they are most similar, thereby achieving the pattern completion that is typical of Gestalt phenomena.

Extracting Chords and Keys

If the input units represent pitch classes, and if their activation decays gradually following the offset of the events that activate them, then units in the output layer will acquire the properties of chord and key detectors (Bharucha 1987, 1991, 1992). This happens in the following way. As each pitch event occurs in a musical sequence, the corresponding input unit is activated. Since this activation takes time to decay, the layer of input units serves as a memory of which pitch classes have occurred and to what extent. This memory essentially serves as a profile of the key (Krumhansl 1990).

Through self-organization, units in the output layer become detectors for entire patterns in the input layer, i.e., they become detectors for keys. If a sequence of tones does not contain all the diatonic tones, or if it does not conform to any of the keys the network has learned, then the activation of the output layer will reflect the keys to which the sequence is most similar.

If the rate of decay of activation at the input layer is made more rapid, then the input layer will represent tonal patterns over shorter temporal windows, for example, chords. In this case, units in the output layer become detectors for chords.

In the case of both keys and chords, the network will exhibit pattern-completion behavior. If activation is allowed to flow back from output units to input units, then pitch classes that are missing from a sequence but are typical of the key or chord will be activated through top-down processing (Bharucha 1987). In other words, a listener may think he or she heard a pitch that did not actually occur, if that pitch typically occurs in that context.

Extracting Pitch

The most interesting pattern-completion phenomenon in pitch perception is the pitch that is heard when the fundamental frequency is missing. Since the early 1970s the prevailing view has been that the missing fundamental is a result of a pattern processing system that fills out the incomplete harmonic spectrum (Goldstein 1973; Terhardt 1972, 1974, 1979; Wightman 1973). Early theories that explained the phenomenon in terms of peripheral mechanisms such as distortion (e.g., Helmholtz 1885/1954) have been rejected by at least three classes of evidence: (1) the missing fundamental cannot be masked by noise in the fundamental's critical band (Licklider 1954); (2) it can be induced by dichotic presentation of subsets of harmonics (Houtsma and Goldstein 1972); and (3) the synthetic pitch sometimes corresponds to neither the fundamental nor other distortion products in the case of pitch shifts (de Boer 1956; Hermann 1912; Schouten et al. 1962).

The models of Goldstein (1973) and Wightman (1973) give a compelling account of aspects of pitch perception, but assume a mechanism for either finding the fundamental or performing Fourier transformations. Terhardt's (1974, 1979) model is a substantial advance because it postulates that this knowledge is learned. Terhardt suggested that we develop harmonic templates from our exposure to speech (vowels, in particular).

According to Terhardt's theory, a signal is matched to learned harmonic templates. The lowest component of each template that produces a match is heard as a virtual pitch. The missing fundamental is heard as a virtual pitch because the upper harmonics are sufficient to produce a strong match to the template of the same fundamental. Virtual pitches other than the fundamental are also generated because of weaker matches. These tend to be subharmonics.

Terhardt (1974) proposed a learning circuit in which the lowest spectral component is extracted and feeds into a pathway that generates a virtual pitch at that frequency. The simultaneous occurrence of the lowest spectral component with higher components lowers resistances connecting the higher components to the same pathway.

A self-organizing neural net can achieve the same result without presupposing a privileged role for the lowest component. Furthermore, it has cognitive generality beyond pitch perception.

In the self-organizing model, units in the input layer layer are modeled after frequency-tuned neurons found in the cochlear nerve (Rose et al. 1971). Units in the output layer develop tuning characteristics through self-organization following experience. The first layer represents the spectrum of frequencies and the second layer represents pitch. The network is presented with harmonic spectra over the speech range, and through self-organization the weights feeding into the pitch units adapt themselves to the harmonic patterns. The weights feeding into each pitch unit take on values that are proportional to the activity levels of the frequency units to which they are connected. In other words, the weight vector feeding into each pitch unit becomes similar to the harmonic pattern that is heard with that pitch. Since the fundamental is the strongest frequency in most harmonic stimuli, the strongest weight will be from the fundamental frequency to the corresponding pitch unit.

Following learning, the network is presented with tones missing their fundamentals. In spite of the absence of the fundamental frequency, the pitch corresponding to the fundamental emerges as the strongest at the pitch layer. This is because the remaining harmonics are sufficient to make the spectrum more similar to the original stimulus with the corresponding fundamental than to any other original stimulus. The pitch layer shows activity levels that reflect relative similarity. Consequently, pitches corresponding to harmonics and subharmonics are also activated.

Conclusion

The same process, neural self-organization, can account for pattern recognition phenomena at many levels. This approach has the promise of unifying the explanations offered for traditional psychoacoustic phenomena as well as more recently studied cognitive phenomena.

References

Bharucha JJ (1987) Music cognition and perceptual facilitation: a connectionist framework. Music Percept 5:1-30

Bharucha JJ (1991) Pitch, harmony, and neural nets: a psychological perspective. In: Todd P Loy G (eds) Connectionism and Music. MIT Press, Cambrige

Bharucha JJ (1992) Tonality and learnability. In: Jones MR, Hollerin S (eds) Cognitive bases of musical communication. American Psychological Association, Washington

De Boer E (1956) On the residue in hearing. PhD thesis, University of Amsterdam

Deutsch D (1969) Music recognition. Psychol Rev 76:300-307

Fukushima K (1975) Cognitron: a self-organizing multilayered neural network. Biol Cybern 20: 121-136

Goldstein JL (1973) An optimum processor theory for the central formation of the pitch of complex tones. Acoust Soc Am 54:1496-1516

Gjerdingen RO (1989) Using connectionist models to explore complex musical patterns. Comput Music J 13(3):67-75

Gjerdingen RO (1990) Categorization of musical patterns by self-organizing neuronlike networks. Music Percept 8:67-91

Grossberg S (1970) Some networks that can learn, remember, and reproduce any number of complicated space-time patterns. Stud Appl Math 49:135-166

Grossberg S (1972) Neural expectation: cerebellar and retinal analogs of cells fired by learnable or unlearned pattern classes. Kybernetic 10:49-57

Grossberg S (1976) Adaptive pattern classification and universal recoding, I. Parallel development and coding of neural feature detectors. Biol Cybern 23: 121-134

Hebb DO (1949) The organization of behavior. Wiley, New York

Helmholtz H (1954) On the sensation of tone. Dover, New York: (Original1885)

Hermann L (1912) Neue Versuche zur Frage der Unterbrechungstone. Arch gesamte Physiol 146:249-294

Houtsma AJM Goldstein JL (1972) The central origin of the pitch of complex tones: evidence from music interval recognition. Acoust Soc Am 51:520-529

Kohonen T (1984) Self-organization and associative memory. Springer, Berlin Heidelber New York

Krumhansl CL (1990) Cognitive foundations of musical Pitch.Oxford University Press, Oxford

Licklider JCR (1954) Periodicity pitch and place pitch. J Acoust Soc Am 26: 945

Pitts W McCulloch WS (1947) How we know universals: the perception of auditory and visual forms. Bull Math Biophys 9:127-147

Rose JE, Hind JE, Anderson DJ, Brugge JF (1971) Some effects of stimulus intensity on response of auditory nerve fibers in the squirrel monkey. J Neurophysiol 34:685-699

Rumelhart DE, Zipser D (1985) Feature discovery by competitive learning. Cogn Sci 9:75-112

Schouten JF, Ritsma RJ, Cardozo BL (1962) Pitch of the residue. J Acoust Soc Am 34:1418-1424

Terhardt E (1972) Zur Tonhöhenwahrnehmung von Klängen. Acoustica 26:173-199

Terhardt E (1974) Pitch, consonance, and harmony. J Acoust Soc Am 55:1061-1069

Terhardt E (1979) Calculating virtual pitch. Hear Res 1:155-182

Von der Malsberg C (1973) Self-organizing of orientation sensitive cells in the striate cortex. Kybernetic 14:85-100

Wightman FL (1973) The pattern-transformation model of pitch. J Acoust Soc Am 54:407-416

What Does the Visual Information Contained in Music Performances Offer the Observer? Some Preliminary Thoughts

Jane W. Davidson

Introduction

Most psychological investigations of music performances have focused on the musical sound, largely ignoring the role of body movements. Exception is found in the writings of Gabrielsson (1985), Jackendoff (1988), and Todd (in press). It is perhaps the work by Shaffer (1980, 1982, 1984), however, which has provided some of the most systematic study of the performer's movements. Although the work takes a motor programming perspective and examines only the durations of key presses in piano performances, it demonstrates that at least part of the pianist's body - in this case the fingers - physically shapes the expressive profile of a performance. The work reveals that this expression has two major components: (1) timing variations away from the metric pulse of the musical rhythm that are inevitable features of motor programming; (2) other timing profile variations which are flexible in note production. These flexible timing features are themselves integrated into the motor program, but are timing effects specifically intended to enhance the musical effect.

The use of the body to enhance expressive effect has been widely discussed in the anecdotal literature on performance. Indeed, Balliot's treatise on violin playing *L'art du violon* of 1834 (cf. Stowell 1985, p. 269) suggests that different tempi should naturally produce different types of physical execution. For example, Balliot remarks that the Adagio tempo results in "more ample movements" than the Allegro where notes are "tossed off", whereas in Presto there is "great physical abandon". Although these points illustrate that movements are essential in the production of an expressive performance, it is also important to note that live music performance is a social communication, which like any other human encounter, presents the perceiver's visual system with information. Indeed, as perceptual verification of this, Schumann's description of the performances given by Liszt reveals how effective this transmission of expression to the audience through body movement can be:

> Within a few seconds tenderness, boldness, exquisiteness, wildness succeed one another; the instrument glows and flashes under the master's

hands... he must be heard and seen; for if Liszt played behind a screen, a great deal of poetry would be lost (An excerpt taken from R. Schumann's *On Music and Musicians* quoted in Morgenstern 1956, p. 155).

Of course, Liszt's performances did not suit everyone's musical taste: Glinka referred to him as the "exaggerator of nuance" (Morgenstern 1956). However, whatever the evaluation of Liszt's style, this example serves to illustrate that body movement provides valuable information about his state of mind and his involvement with the music. The illustration of Liszt in Fig. 1 shows that flourishes of the body are evidently intended to have meaning and communicative force for the audience. Although nowadays there is a diminished enthusiasm for this kind of extrovert performance, professional performers nonetheless talk about "projecting" the

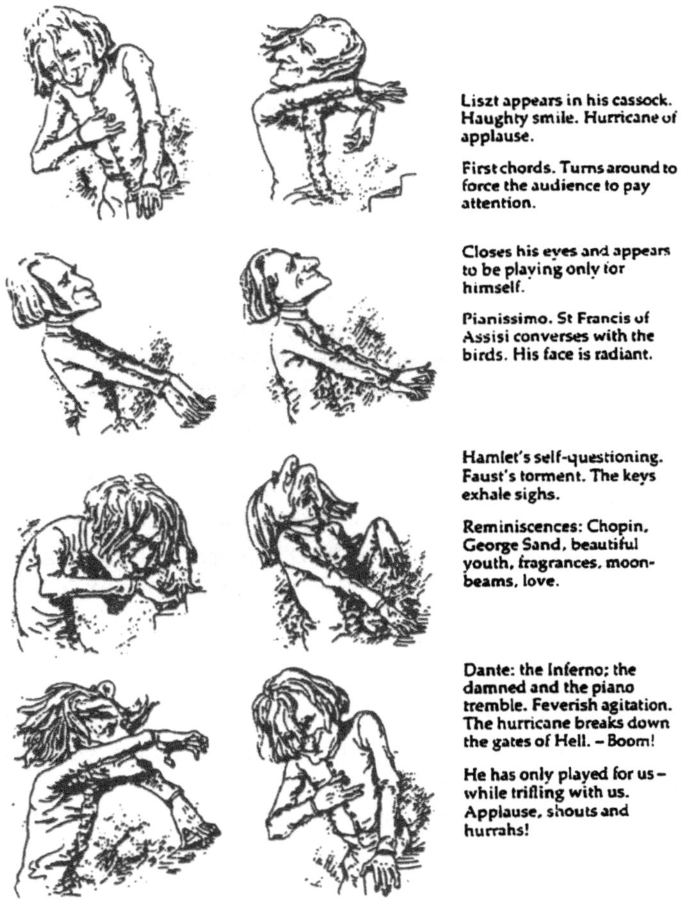

Fig. 1. Caricatures of Liszt's performances by Janko (1873; reproduced in Sachs, 1982)

performance through the body; Baker (1982) claims that a performer can only communicate to the audience if she is prepared to allow the sounds to live and breathe through the body.

Studies by Clynes (1980) and Clynes and Nettheim (1982) have suggested that visual information is of the same expressive type as sound. Their work refers to movements and sounds conveying equivalent *sentic form* (characteristic dynamic expressive forms of specific emotional qualities). There conclusions are based upon using the stimuli of seven different hand movements which Clynes and Nettheim considered to be analogous to seven emotional qualities (anger, hate, grief, love, sex, joy, and reverence). When they asked white urban Australian and American observers and Aboriginals living in the Australian bush to identify the movements they discovered that the appropriate emotional qualities were ascribed to each by all groups. Similarly, when single musical sounds were given amplitude profiles similar to those of the hand movement patterns, they were identified as containing the same emotional qualities as the pressure movement from which the sound was derived. Whilst there has been little properly controlled empirical validation of these ideas, Clynes and Nettheim suggest that the intimate link between sound and movement has perceptual as well as production significance.

The fundamental concern of this chapter, therefore, is to explore the perceptual information contained in the visual appearance - that is to say the performer's body movements - of music performances. The discussion which partly draws on already published material (Davidson 1993) will be based around studies in which a variety of explicit performance interpretations are examined.

What makes different performance interpretations of the same piece particularly interesting is that whilst it is clearly necessary for the performer to obey the instructions of a score, the performer is also expected to have a particular approach to the piece and therefore the presentation of the music. Thus it should be possible to visually distinguish the different interpretations. An empirical basis for this hypothesis can be found in Runeson and Frykholm's (1983) investigation of the different types of expressive information that are available in total body movement tasks like walking, throwing and lifting. In the case of lifting, for example, when shown film of an actor picking up a box, observers were able to detect the box weight and how much that weight differed from the lifter's expectation about the weight. Furthermore, attempts to give false information about the box weight were detected by the onlookers.

The significance of these findings is that covert mental dispositions become specified in movement and are detectable to observers (in this case, the box lifter's expectation, deceitful attempt and real weight lift are all specified). Runeson and Frykholm encapsulate this in their proposal that there is a kinematic specification of dynamics. That is, movements are specific to the conditions that generate them.

Empirical Investigations: How Does Body Movement Aid Us in Music Interpretation?

Methodological Questions

Without empirical precedents in psychological enquiries into the visual perception of music performances, it is essential to find a suitable experimental tool. This section describes a way to investigate this question and discusses the results achieved.

Methodologically, visual perception research is particularly tricky given the large amount of contextual and situational information in any visual scene. This means that it is often difficult to know whether the perceiver is focusing on and being influenced by the performance space, the performer's dress, hair style or specific features of body structures. Therefore, in order to focus the observers on the dynamic human movement stimulus material, Johansson's (1973) pioneering "point - light technique" was adapted. The method uses ribbons of glass-bead retroflective tape attached to the major body joints and the head, and spotlights mounted very close to the lens of a camera so that the light hitting the reflective

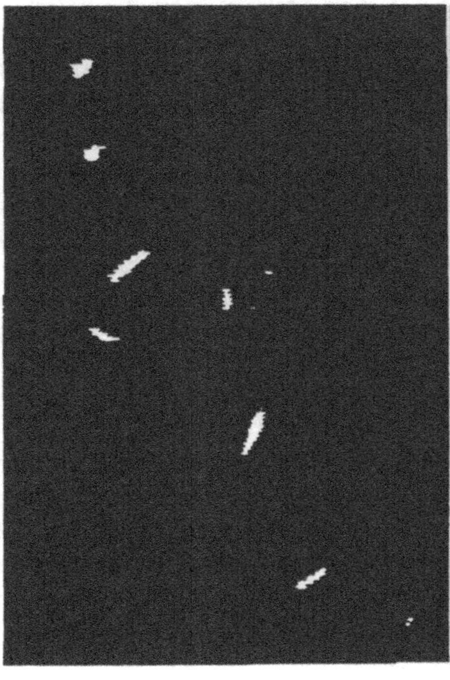

Fig.2a. Snap-shot of a pianist in point-light conditions

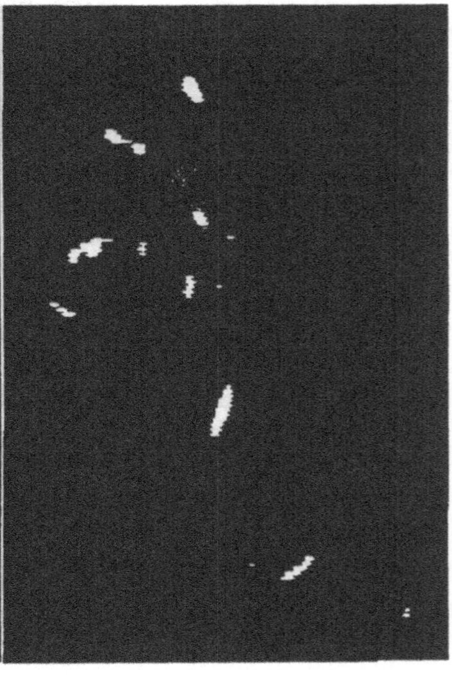

Fig.2b. Snap-shot of a pianist in point-light conditions.

ribbons is reflected back into the camera. The result is to have a stimulus based solely on movement. This is created by an extremely high brightness contrast between the ribbons and the background. This is viewed in playback by adjusting the brightness and contrast control of the monitor to produce a supercontrast between the ribbons and the background. Thus, only the ribbons which are seen as spots of light can be viewed on the monitor. Still examples of the effect can be found in fig. 2a,b which show two snapshots from a pianist's performance in the pointlight condition.

The method had already yielded impressive results in simple performances such as the box lifting mentioned above (Runeson and Frykholm 1983) and artistic performance in Scully's (1986) work on gymnastics performances. In addition to Runeson and Frykholm's discovery that observers were able to detect the differences in the performer's intentions, Scully found that discriminations between technical and aesthetic qualities of a balance beam routine could be made.

The Studies

In this set of studies, 21 musician observers were asked to consider a series of solo performances given by four violinists and 34 musician observers to consider the performances of a pianist. The performers had different expressive intentions - that is, they were asked to perform the same piece of music in particular manners. The manners chosen were genuinely performable. One of these was consistent with a public performance - the "projected" performance - and the other two ("deadpan" and "exaggerated") were equivalent to teaching methods. In the deadpan performance, the performer was asked to produce the music in a purely technical manner. This device is used by teachers to encourage students to focus on the technical aspects of performance. In the "exaggerated" manner, the performer was asked to play the piece in such a way that all aspects of the music's expressive features would be overstated. This teaching device is commonly used to encourage an awareness of the music's expressive features. No specific instructions were given about how to achieve these differences since the instructions were viewed as a simple method to get the performers to approach the music with different intentions and to see if these differences were translated into movement differences.

The observers were presented with these manners in three modes: the visual point - light stimuli alone, the sound alone, and the combined visual and sound stimuli. This was done to investigate whether the manners were conveyed in each stimulus mode.

Results

First of all, the studies illustrated that there is sufficient perceptual information contained in visual information of kinematics alone to permit the identification of performance manner. Therefore, the point - light technique enabled the empirical establishment that visual information provides accurate information about performance intention. In addition, the results revealed that although vision, sound, and sound and vision modes all showed differences between manners, it was vision mode which provided the greatest scoring difference between the manner scores. That is, observers reported their perceptions by rating the intention of the performances on a bipolar deadpan - exaggerated rating scale, and in the vision-only mode deadpan manner was given the lowest deadpan rating and exaggerated was given the highest exaggerated rating. Furthermore, vision was the only mode to significantly differentiate between projected and exaggerated manners. Together, these results suggest that vision most strongly conveyed the differences between the three performance manners and therefore suggested that vision is the most "effective" indicator of manner.

This final result raises a number of issues which require detailed consideration. Before this takes place, however, there is one empirical follow-up which was pursued and will be mentioned here, since its results have an effect on the interpretation of the previously described study.

Observer Experience and Its Effect

This work is an unpublished study which was designed to examine musician and nonmusician perceptual responses to independent visual and sound stimuli. Given the results of the study described above, it was hypothesised that without former experience of classical music sound stimuli, the nonmusicians would be more dependent on visual stimuli for information about performance manner than the musicians.

The study used point - light stimuli of five solo performers in sound and vision modes playing in the deadpan, projected, and exaggerated manners similar to the study described above. Ten musicians and nonmusicians (the nonmusicians were adults who had never attended a classical music concert and who did not own or consciously listen to classical music recordings) were asked to rate performances in the different sensory modalities.

It was discovered that as in the previously described study, musicians could discriminate between the deadpan, projected and exaggerated manners in the sound, vision and sound and vision modalities; however, the nonmusicians found the task more difficult and could only reliably discriminate in the vision-only mode.

The numbers of observers used in the study is too small to draw firm conclusions. However, as a preliminary investigation it suggests that for the nonmusician, visual stimulus mode may be the main source of information about the expressive intention of the performer.

What Can Be Learned from the Investigation of the Visual Information Contained in Music Performances?

Whilst music itself is an aural phenomenon, it seems that the information contained in the performer's body movements provide highly effective indications of what the performer is attempting to achieve. Indeed, for the observer who is not skilled at listening to classical music, the visual information may inform their perceptions. Therefore, it seems strange that so little attention has been paid to this visual information in the study of the music performance. One explanation for this is the domination in Western culture of the aural recording. However, simply because the aural recording is the most common way of experiencing music, it does not necessarily follow that the aural recording is the most effective way of experiencing music.

One interesting contemporary anecdotal follow up to this point can be found by considering the career of the pianist Glenn Gould. Having observed video recordings of Gould from early and late in his career, F. Delalande (1990, unpublished paper presented at the Second International Colloquium on the Psychology of Music at Ravello) noted a marked difference in the pianist's movements. Early in his career, Gould performed in public, and his performances show a fluency, yet a degree of unpredictability in his movements. His later performances which were restricted to the recording studio reveal fixed, repetitive series of movements which recur in a predictable fashion. It must be assumed that years of performing without the presence of an audience affected his performance presentation. It is feasible that in the latter performances he was not motivated by communicative socially constructed movement concerns at all, and his movements took on a purely structural character. So the communicative context can, in turn, feed back into the musician's performance.

This last point brings with it many issues which must be brought to the attention of the musician. Amongst these are questions about how the performer should work on a piece of music for live performance, bearing in mind that there is no formal emphasis on the movements of performance, and the player spends many hours alone without feedback about what is being communicated in her performance. This seems ironic given that live musical performance is a social activity and perhaps more time ought to be spent in social situations when practising for concerts.

Of course, Delalande's work and the findings discussed in this chapter raise questions about the way in which performances are experienced by the observer. For instance, it leads one to wonder about the expressive impact of music which was originally conceived for live performance and how it is perceived when it is presented as an aural recording alone. Returning to the earlier examples referring to Liszt, it must be asked whether sound only is an adequate way to experience his compositions. Obviously, expressive affect differs depending on the pervading cultural aesthetic, and the melodramatic style of Liszt's epoch does not fit with the current model. However, it may be that the absence of movement stimuli is significant to the perception and therefore the experience of Liszt's music as it was conceived by him. This may be a more important issue than whether the musical sound is presented as a stylistic replica of his own performances.

It is important to understand why visual information is a stronger indicator of manner than sound. At the level of the nonmusician, the answer may lie in the fact that non-musicians are not skilled at listening to differences in expression between performances, whereas they are skilled observers of expressive differences in movement (Runeson and Frykholm 1983). It has to be pointed out though, that music performances were also novel visual experiences to all the nonmusician observers in the study. This does not, however, provide a suitable explanation for why musicians found the visual information stronger than sound. One possible explanation is that musical sound alone is intrinsically more expressive than musical movement. That is, the actual characteristics of music (tonal and rhythmic) make it expressive. Therefore, there is never an absolute baseline of no expression. For movement, whilst it is essential to move in order to produce the sounds, the expressive baseline may be lower since not all of the intrinsic expressivity of the music is reflected in the physical act of producing the performance. For instance, the tonal features of the music cannot be seen, though rhythmic features may be. This explanation accounts for the deadpan result alone. Yet, a similar model can be drawn on to explain the higher exaggerated rating for vision too. Indeed, it could be that the range from 0 to maximal expressivity is greater in movement than sound. For, whilst sound may have a higher intrinsic expressivity, it may have a lower maximal expressivity level within the constraints of certain musical features. These constraints may include: the maximum volume at which an instrument can be played and the maximal deviations in tempo and rhythm away from the compositional indications that are possible without changing the musical meaning of these indications. In movement there may be more freedom, and therefore, greater potential for a higher maximal level. However, this question needs to be addressed in detail in further empirical investigations. Such studies are essential if the role of movement in the perception of live performances is to be fully understood.

In order to address this issue and others, future research could begin by asking what part of the movement pattern specifies the expressive information. In the specific case of Davidson (1993) the three performance manners need to be examined to investigate what differentiates and unifies them.

Clearly, whilst research on the contribution of body movement to musical expression and its perception has begun, there are many avenues of inquiry still to be pursued.

References

Baker J (1982) Full circle: an autobiographical journal. MacRae, London

Clynes M (1980) The communication of emotion: theory of sentics. In: Plutchik R, Kellerman H (eds) Theories of emotion. Academic, New York

Clynes M, Nettheim N (1982) The living quality of music: neutrobiotic patterns of communicating feeling. In: Clynes M (ed) Music, mind and brain: the neuropsychology of music. Plenum, New York, pp 47-82

Davidson JW (1993) Visual perception of performance manner in the movements of solo musicians. Psychology of Music, 21:103-113
Gabrielsson A (1985) Interplay between analysis and synthesis in studies of music performance and music experience. Music Percept 3(1):59-86
Jackendoff R (1988) Consciousness and the computational mind. MIT Press, Cambridge Mass
Johansson G (1973) Visual perception of biological motion and a model for its analysis. Percept Psychophy 14:201-211
Morgenstern S (1956) Composers on music. Faber and Faber, London
Runeson S, Frykholm G (1983) Kinematic specification of dynamics as an informational basis for person-and-action perception: expectations, gender, recognition, and deceptive intention. J Exp Psychol Gen 112:585-615
Sachs H (1982) Virtuoso. Thames and Hudson, London
Scully DM (1986) Visual perception of technical execution and aesthetic quality in biological motion. Hum Mov Sci 5:185-206
Shaffer LH (1980) Analyzing piano performance: a study of concert pianists. In: Stelmach GE, Vroon PA (eds) Tutorials in motor behaviour. North Holland,Amsterdam, pp 443-456
Shaffer LH (1981) Performance of Chopin, Bach and Bartok: Studies in motor programming. Cog Psychol 13:326-376
Shaffer LH (1982) Rhythm and timing in skill. Psychol Rev 89:109-122
Shaffer LH (1984) Timing in solo and duet piano performances. Q J Exp Psychol 36 [A]:577-595
Stowell R (1985) Violin technique and performance practice in the late eighteenth and early nineteenth centuries. Cambridge University Press, Cambridge
Todd NPMcA (in press) The kinematics of musical expression. J Acoust Soc Am (in press)

Tone Discrimination in Patients with Temporal Lobe Lesions

Heinz-Gregor Wieser and Elke Wittlieb-Verpoort

Introduction

Within the framework of working memory, subsidiary systems have been investigated for the retention of verbal and nonverbal material. While the majority of this research has been concerned with the verbal short-term store (see Baddeley 1986 for a review), the storage of nonverbal material has also been investigated with different types of stimuli. A visual - spatial short-term store was proposed for maintaining simple visual stimuli and complex spatial material (Phillips and Christie 1977; Baddeley and Lieberman 1980; Logie 1986). This temporary store was found to be susceptible to disruption by concurrent spatial processing. Eye movements were assumed to serve as control processes rehearsing and maintaining the visual image. Nonverbal auditory material, such as tones, has mostly been investigated in the form of sequences either at a perceptual level in terms of pitch discrimination (Moore and Rosen 1979; Patterson et al. 1983) or with respect to the cognitive representation of tonal relationships (Deutsch 1984; Dowling and Harwood 1986). Little neuropsychological work has been done on the storage and rehearsal of these auditory stimuli. Early studies (Kimura 1961a,b; Shankweiler 1966) reported a double dissociation for the site of lesion (left or right temporal) and type of material (digits or melodies). Patients with left epileptogenic lesions showed more errors on the dichotic digits test and patients with similar right-sided lesions were impaired in the recognition of melodies. Bever and Chiarello (1974), however, suggested that the side of processing melodies was related more to different processing style of the two hemispheres than to their specialization for verbal or nonverbal material. In their experiments, musicians demonstrated a right-ear advantage for the recognition of dichotically presented melodies, whereas untrained listeners showed the usual left-ear (right hemisphere) advantage. Assumung a serial, analytic processing style for the left and a holistic, unsegmented style for the right hemisphere, Bever and Chiarello attributed peformance differences to subjects' varying familiarity with musical structures. Trained musicians were thought to apply their knowledge of tonality to the required task, thus analyzing the sequences in small segments, whereas naive listeners were thought to treat the sequences as unanalyzed wholes. The group differences in hemispheric processing of melodies were supported by EEG recordings (Hirshkowitz et al. 1978), and they were reported to correlate with subjects' musical aptitude (Gaede et al. 1978). Musicians' superiority in recognizing melodic chords was, however, confined to those conforming with the rules of Western tonal harmony (Beal 1985). Nonmusicians seemed to base their judgments on the acoustic properties of chords, thus being less affected by chords which did not fit tonal relations. When applying concepts of the working memory model to the processing of melodies, there was evidence for a functional overlap of mechanisms for processing verbal and nonverbal "semantic" material (melodies). Logie and Edworthy (1986) observed reduced melody recognition with a concurrent homophone judgment or an articulatory suppression task, but little effect with visual matching of nonalphabetic symbols. The authors concluded that besides an articulatory code for storing verbal material, "subvocal singing" was to maintain the acoustic image of melodies.

The cited studies revealed that the more a tone sequence implied a learned tonal structure, and the greater subjects' abilities were to analyze it according to tonal rules, the more likely these processes were mediated by the dominant/left hemisphere. The nonverbal task of the present study, however, was to address short-term memory functions of the nondominant/right hemisphere. Therefore, a tone discrimination task was chosen which required pitch discrimination of pure tones. These stimuli were to avoid "semantic" tonal relationships among tones as assumed for melodic tone series. The retention of pitch has extensively been studied by Deutsch (1970), who had subjects compare the pitch of two tones and were asked to judge whether the pitch of the tones were same or different. While interpolated digits only produced minimal performance decrements, even when their recall was required, interpolated tones were found to produce a substantial increase in errors of pitch recognition. As the error rate varied systematically with the degree of similarity between the test tone and a critical interpolated tone, Deutsch (1972) suggested a specialized system for the processing of pitch with tones interacting within this system. Because of this interaction among pitches, the difficulty levels of the present tone discrimination task were determined by a series of pretests (Wittlieb-Verpoort and Perret 1989) which investigated frequencies and number of tones in various discrimination tasks. As impaired retention of nonverbal material had been reported for various stimuli with right temporal lobe lesions (Kimura 1963: nonsense figures; Milner 1965, 1968: mazes, faces), damage in this region was assumed to produce pronounced impairment in the discrimination of pure tones with greater memory load.

Methods

Subjects

A total of 52 brain-damaged patients and 20 control subjects were tested. Four patients with atypical speech representation (as measured by Oldfield Handedness

Table 1. Demographic data of subject groups

	Subject groups					
	Temporal		Frontal		Control Subjects	Statistics
	Left ($n = 11$)	Right ($n = 13$)	Left ($n = 10$)	Right ($n = 9$)	($n = 20$)	
Sex						
Male	5	5	6	7	10	Chi - (4) =
Female	6	8	4	2	10	3.84
						$p = .43$ [a]
Age						
Mean	33.2	30.9	39.5	38.4	26.5	$F(3,39) =$
Median	32	33	39	35	26	1.38
Range	19-54	18-44	24-54	24-63	21-35	$p = .26$ [b]
Educational background [c]						
9 years	2	3	2	0	0	
13 years	7	9	5	6	0	
> 13 years	2	1	3	3	20	

[a] A valid chi-square test should not have more than 20% of cells with expected counts less than 5 and none with counts below 1 (Cochran 1954). In this test 30% of cells had expected counts less than 5 so that assumptions about the distribution of data should be treated with caution.
[b] One-way analysis of variance (F-test) for patient groups. Control subjects were significantly younger than patient groups [$F(4,58) = 13.6, p < .001$].
[c] Cell statistics did not allow valid chi-square test.

Inventory or the Intracarotid Amytal Test) were eventually excluded from the analysis. Three further patients did not finish testing due to fatigue and two others scored below average on one of the control tests (Token Test). These patients were not included into the study. The four groups consisted of subjects with either left ($n = 11$) or right ($n = 13$) medial temporal lesions (LT or RT patients, respectively) and left ($n = 10$) or right ($n = 9$) frontal lesions (LF or RF patients, respectively). LT and RF patients served as control groups for RT patients with respect to opposite- and same-hemispheric processing, respectively. LF patients represented a brain-damaged control group whose impairment was not thought essential in solving the tone discrimination task. Six patients with medial temporal lesions had been operated

Table 2. Etiology of brain damage

	Temporal		Frontal		Statistics
	Left	Right	Left	Right	
Months after surgery					
Mean	15.5	11.5	29.6	25.6	$F(3,39) =$
Median	10	7	18	21	1.49
Range	0-68	0-34	0-69	0-102	$p = .23$
Months of illness [a]					
Mean	94.9	99.9	48.2	36.8	$F(3,39) =$
Median	47	58	48	25	1.45
Range	5-312	7-420	11-120	6-108	$p = .25$
Etiology					b
Pre/post [c]	3/8	2/11	1/9	1/8	
Tumors	3	4	7	8	
CVA	4	3	3	1	
PSAHE	1	5	-	-	
Seizures [d]	3	1	-	-	

CVA, cerebral vascular accidents; PSAHE, palliative selective amygdalohippocampectomy

[a] The first onset of psychomotor seizures was counted if reported at least once a month for five consecutive years.
[b] Cell statistics did not allow chi-square test.
[c] Tone discrimination testing was before (pre) or after (post) neurosurgical intervention.
[d] Psychomotor seizures.

for complex partial seizure. This type of surgery involves the selective amygdalo-hippocampectomy sparing the dorsolateral and laterobasal portion of the temporal lobe (Wieser and Yasargil 1982). Other patients had been operated on for tumors or cerebral vascular accidents (CVA) (medial temporal: $n = 13$, frontal: $n = 17$). Clinical localization of lesions was assisted by EEG, magnetic resonance tomography, or operation. These patients had no further neurological signs. Patient groups were matched for age, sex, and educational background. Control subjects (CT) were students and members of the hospital staff. The sociodemographic data of the groups and the etiology of brain damage are presented in Tables 1 and 2.

Test Materials

Three levels of difficulty were defined for this task, reflecting an adjusted task demand for brain-damaged patients (Wittlieb-Verpoort 1989). Task 1 served as a simple "control task" to ensure that subjects were able to discriminate pure tones with minimal memory load. In task 1 (TD), target tones of 500 Hz and distractor tones of 600 Hz were given. The more difficult tone series of task 2 (DTTTD) contained three different target frequencies (400, 500, 600 Hz) and two distractor frequencies (300 and 700 Hz), the latter being at the beginning and at the end of the frequency range. Task 3 (DTDT) comprised two target frequencies of 500 and 700 Hz and two distractor frequencies of 400 and 600 Hz. In the test series of each task level, target and distractor tones were presented at random (Fig. 1).

Fig. 1. Levels of task difficulty

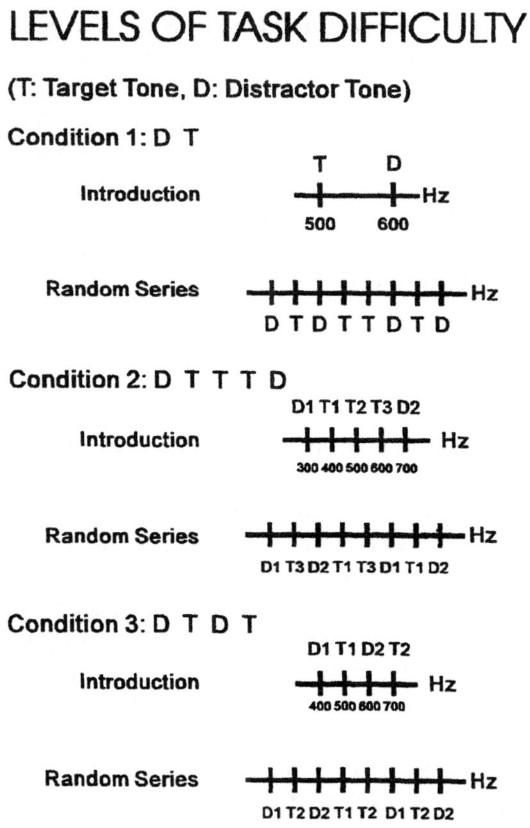

T, target tone; D, distractor tone

The arrangement and detectability of pitches was given preference over the absolute number of targets in determining levels of task difficulty. The tone series of tasks 1, 2, and 3 consisted of, in order of presentation, 14, 26, 20 or eight tones and lasted for either 34, 64, 49, or 19 respectively. The tones were presented for 1.5 s with an interstimulus interval of 1 s to allow for subjects' response when identifying signals als targets (Munson, 1947; Stevens and Davis 1938). The tones were produced by a two-channel sinus tone generator (EL Elektronik, Lehnherr, Basel, Switzerland) and were presented binaurally via headphones. Subject and experimenter both wore stereophonic headphones permitting the latter to monitor the experiment. Tone signals were presented within a range of 50 to 75 dB/sound pressure level (Bruel and Kjoer, Copenhagen, 2606), allowing subjects individual adjustments in intensity when the first set of tones was introduced. Although this adjustment did not account for differences in perceived pitch across subjects, pitch intensity of the different tones varied by only 10 dB, +/-2 dB, depending on the intensity level initially chosen by each subject. As changes in perceived pitch appeared to be relatively minimal (Robinson and Dadson 1957), equal loudness contours were not formally controlled.

In addition to the experimental tests, a series of neuropsychological tests was administered to subjects in order to control for the specific functional deficits associated with the cicumscribed lesions. These tests were part of the routine testing procedure applied at the Neuropsychological Institute in Zurich and served as a global assessment of patients' cognitive level of functioning as well as a rough screening for multile or atypical dysfunctions. The series included the Token Test (De Renzi and Vignolo 1962), Digit Span (Wechsler 1964) and Block-tapping Test (Corsi 1972), the Picture Arrangement Test (McFie and Thompson 1972) and a controlled word association test.

Procedure

At each level of task difficulty, the tones were first introduced and defined as targets or distractors to subjects who were to memorize the targets and identify them verbally in the subsequent tone series. In task 2 and 3 with more than one target frequency, subjects were asked to respond to any target, irrespective of its actual frequency. Each task was first practiced with a series of ten tones to familiarize subjects with the pace and pitch of tones. If necessary, the practice trial was repeated twice. Then, the frequency arrangement of tones was given once again, before the first test series started with a delay of 2 s. The experimenter compared subjects' responses with the correct answers for each tone. In practice and test series, subjects were corrected if target tones were missed or distractors labeled as targets. The experimenter identified the next tone as either target or distractor in order to reorient subjects in frequency and to avoid further dependent errors. The three task conditions were interspersed with the control tests.

Although these tasks did not employ melodies, subjects were questioned about their musical training or activities to account for possible facilitating effects. A score of one was given to subjects for less than 1 year of musical activity (e.g., singing, practiced instruments) longer than 5 years before the time of testing. Subjects received a score of three for current musical practice for 5 years or longer. Any other musical activity, e.g., either done for a long period of time more than 5 years ago or only occasionally performed, was scored as two.

Design and Statistical Methods

Statistical analysis was calculated for a randomized block design (Kirk 1968). The data were analyzed by analyses of variance (ANOVA) and calculated by the Biomedical Statistical Program (BMDP4V) for repeated measures. As a priori hypotheses had been stated about subjects' performance on the task, cell means were compared by planned orthogonal and nonorthogonal pairwise comparisons. The probability of the type I error was set for the collection of comparisons with common cell means, but was kept for the individual comparisons if contrasts were independent (Kirk 1968). Differences among means are reported if the alpha level was less than 0.10. The small sample size of the brain-damaged subject groups did not permit conclusive assumptions about the symmetry of the dependent measures (Winer 1971) of the within n-factors. Therefore, the more conservative F-test by Geisser and Greenhouse (1958) with reduced degrees of freedoms was taken as statistical basis.

Results

As there were different probabilities of making errors in the four tone series, the analyses were based on error ratios (number of errors/number of tones). Figure 2 shows an increase of ratios with greater task difficulty.

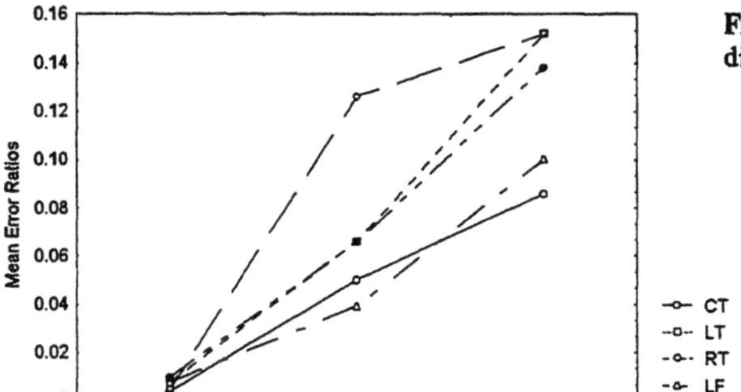

Fig. 2. Mean error ratios of tone discrimination task

RT, right temporal; LT, left temporal, RF, right frontal; LF, left frontal; Ct, healthy controls

Only RT patients exceeded the 20% error ratio in the first tone series (14 tones) of task 3 ($M = 22\%$). RT patients showed higher mean ratio than any other group; these differences were most clearly observed in task 2 (DTTTD). The error ratios were submitted to a three-factorial ANOVA with five subject groups, three task levels (TD, DTTTD, DTDT) and four series lengths (8, 14, 20, 26 tones) revealing significant differences among subject groups ($F = 3.03$, $p < 0.05$), a significant main effect of task difficulty ($F = 65.4$, $p < 0.001$) and a significant interaction between the two within n-factors ($F = 5.10$, $p < 0.001$). This interaction referred to consistent differences among task levels for the three longest series (14, 20, 26 tones)

The interaction between subject groups and task difficulty approached statistical significance ($F = 1.89$, $p = 0.08$). When analyzing this interaction by pairwise comparisons, there were no significant differences among groups in task 1 (TD) (Table 3).

Table 3. Mean error ratios and standard deviations of tone discrimination task

	CT	LT	RT	LF	RF
Conditions					
T1QU	0.003 (0.015)	0.009 (0.040)	0.006 (0.019)	0.008 (0.025)	0.007 (0.022)
T2QU	0.049 (0.067)	0.065 (0.076)	0.126 (0.129)	0.039 (0.074)	0.066 (0.074)
T3QU	0.085 (0.114)	0.150 (0.121)	0.170 (0.130)	0.098 (0.105)	0.137 (0.111)

CT, control subjects; LT, left medial temporal lesion; RT, right medial temporal lesion; LF, left frontal lesion; RF, right frontal lesion

Table 4. Correlations between musical practice and errors

	CT	LT	RT	LF	RF
Task levels					
1	0.04	- 0.24	0.07	0.00	0.10
2	- 0.28*	- 0.48*	- 0.05	- 0.33*	- 0.21
3	- 0.38*	- 0.52*	- 0.11	-0.52*	- 0.30

For abbreviations see Table 3.
* $p < .05$

Task 1 showed an overall low mean error ratio ($M = 0.6\%$) across all five groups. In task 2 (DTTTD), the overall mean ratio increased markedly ($M = 6.8\%$). RT patients provided the highest mean ratios ($M = 12.6\%$) with significant differences between this group and either controls ($M = 4.9\%$) or LF patients ($M = 3.9\%$) ($F = 11.3, p = 0.001$; $F = 10.3, p = 0.002$, respectively). While the overall performance deteriorated in task 3 (DTDT) ($M = 12.3\%$), differences between tasks 2 and 3 were less pronounced than between tasks 1 and 2. Although RT patients still showed the highest error ratios ($M = 17\%$), group differences began to diminish as LT and RF patients also made more errors in task 3 ($M = 15\%$, 13.7%, respectively). In contrast, controls and LF patients generally had the lowest mean error ratios in both tasks 2 and 3.

The extent of musical practice tended to be greater in left-hemipheric than right-hemispheric patients (Table 4).

Controls reported the most active musical involvment. Statistical analysis of the distribution, however, was not meaningful as there were too many cells with counts less than 5. The overall correlations (Spearman rank correlation coefficient) between subjects' musical practice and their task performances (mean error) were significant for controls and left-hemispheric patients. Significant negative correlations were obtained in tasks 2 and 3; the coefficients tended to increase with greater task difficulty, except for RT patients. Controls and LT and LF patients showed a clear relationship between greater musical practice and better task performances, whereas the correlations of RT and RF patients were not significant.

Discussion

The monitoring of two tones in task 1 (TD) seemed to have been so easy for all subject groups that no differential effects were obtained. Performance decrements were found in tasks 2 and 3 when three and two target tones were to be identified, respectively. RT patients presented the highest error ratios in these tasks and the most pronounced group differences in task 2. Previous studies (Wittlieb-Verpoort and Perret 1987) had shown that subjects tended to simplify task 2 by memorizing the two distractor instead of the three target pitches. Apparently, controls and LF, RF, and LT patients of the present study also employed this strategy, producing fewer errors in task 2. The markedly increased ratios of RT patients in this task indicates that they might not have recoded tones but tried to memorize the three target pitches individually. Consequently, processing capacity of the working memory must have been exceeded when simultaneously rehearsing three target tones and match their pitches with the tones of the random sequence. If, however, RT patients had indeed recorded tones and were rehearsing the two distractor tones, this strategy implied to selectively respond to the appropriate stimuli. Subjects had to respond to tones of the random sequence (targets) which were different from the ones being rehearsed (distractors), and RT patients might not have been able to produce or suppress the relevant responses to the presented stimuli. In contrast, the arrangement of target pitches in task 3 (DTDT) prevented subjects from recoding tones and require the rehearsal of all four tones to keep oriented within the scale. This task seemed to generally have challenged processing capacity as even controls demonstrated an appreciable increase of errors. The specific problems discussed for RT patients reveal the significance of control processes (Atkinson and Shiffrin 1968; Baddeley 1986) in the maintenance of this auditory, nonverbal material. The particular controlled processing required in this tone discrimination task was obviously impaired with damage to the right temporal lobe. If the recoding of tones is viewed as a reorganization of stimuli, these findings correspond well with the hypothesis discussed by Jones-Gotman (1986). The author reported deficient structure and organization in patients with right mediobasal lesions who were required to reproduce a complex figure after a 40-min delay. She interpreted her results in support of an organizational role of the hippocampus, the structure facilitating the maintenance of stimuli over longer periods of time. Although in the present study an external structure had been given to subjects in terms of the definition of tones (targets, distractors) and by explicit instructions, all except RT patients seemed to have spontaneously reorganized this task. The absence of spontaneous recoding of tones in RT patients would support the organization hypothesis of the hippocampus.

Various subjects had incidentally commented that they had tried to hum the pitches to be remembered. These comments again underline the role of motor programs in prolonging the storage of stimuli. Similar findings have been reported for the retention of melodies by "subvocal" singing (Logie and Edworthy 1986) or for visual spatial material by eye movements (Phillips and Christie 1977). The rehearsal of target pitches, however, seems to have been disrupted by concurrent tone discrimination. Differential effects were particularly found in longer tone series (14, 20, 26 tones), i.e., exceeding a 19-s interval. The findings coincide with the notion of a critical time interval for the short-term storage of information, implying the maintenance of stimuli for a certain period of time even without explicit rehearsal of stimuli and or with concurrent processing.

The inverse relationship between musical practice and error ratios for controls and left-hemispheric patients suggested that these subjects might have taken advantage of their musical practice, e.g., their knowledge about melodic harmonies, to solve the tone discrimination task.

In conclusion, the maintenance of pure tones was found to benefit from spontaneous recoding of stimuli or learned concepts about melodies. It was assumed that patients with right temporal lobe lesions did not apply either of these strategies efficiently, the results supporting the organizational role of this region.

Acknowledgements. The authors are grateful to the Swiss National Science Foundation supporting this research through grant no. 3.832.-.84.

References

Atkinson RC, Shiffrin RM (1968) Human memory: proposed system and its control processes. In: Spence KW, Taylor Spence J (eds) The psychology of learning and motivation, vol. 2, Academic Press, New York, pp 89-195

Baddeley AD (1986) Working memory. Oxford University Press, Oxford

Baddeley AD, Lieberman K (1980) Spatial working memory. In: Nickerson R (ed) Attention and performance VIII. Erlbaum, Hillsdale, pp 521-539

Beal AL (1985) The skill of recognizing musical structures. Memory Cogn. 13:405-412

Bever TG, Chiarello RJ (1974) Cerebral dominance in musicians and nonmusicians. Science 185:537-539

BMDP (1987) Biomedical computer programs. University of California Press, Los Angeles

Corsi PM (1972 Human memory and the medial temporal region of the brain. PhD thesis, McGill University, Montreal

De Renzi E, Vignolo LA (1962) The Token test: a sensitive test to detect disturbances in aphasics. Brain 85:665-678

Deutsch D (1970) Tones and numbers: specifity of interference in short-term memory. Science 168:1604-1605

Deutsch D (1972) Effect of repetition of standard and comparison tones on recognition memory for pitch. J Exp Psychol 93: 156-162

Deutsch D (1984) Memory for nonverbal auditory information: a link between behavioral and physiological studies. In: Squire LR, Butters N (eds) Neuropsychology and memory. Guilford, London, pp 45-54

Dowling WJ, Harwood DL (1986) Music cognition. Academic, New York

Gaede SE, Parsons OA, Bertera JH (1978) Hemispheric differences in music perception: aptitude versus experience. Neuropsychologia 16:369-373

Geisser S, Greenhouse SW (1958) An extension of Box' s results on the use of the F-distribution in multivariate analysis. Ann Math Statistics 29:885-891

Hirshkowitz M, Earle J, Paley B (1978) EEG alpha asymmetry in musicians and non-musicians: a study of hemispheric specialization. Neuropsychologia 16:125-128

Jones-Gotman M (1986) Memory for designs: the hippocampal contribution. Neuropsychologia 24:193-203

Kimura D (1961a) Some effects of temporal-lobe damage on auditory perception. Can J Psychol 15:156-165

Kimura D (1961b) Cerebral dominance and the perception of verbal stimuli. Can J Psychol 15:166-171

Kimura D (1963) Right temporal-lobe damage. Arch Neurol 8: 264-271

Kirk RE (1968) Experimental design: procedures for the behavioral sciences. Brooks/Cole, Belmont

Logie RH (1986) Visuo-spatial processing in working memory. Q J Exp Psychol 38:229-247

Logie RH, Edworthy J (1986) Shared mechanisms in the processing of verbal and musical material. In: Russell DG, Marks DF, Richardson JTE (eds), Imagery 2. Proceedings of the 2nd International Imagery Conference, 1985. Human Performance Association, Dunedin, pp 33-37

McFie J, Thompson JA (1972) Picture arrangement: a measure of frontal lobe function? Br J Psychiatry 121:547-552

Milner B (1965) Visually-guided maze learning in man: effects of bilateral hippocampal, bilateral frontal and unilateral cerebral lesions. Neuropsychologia 3:317-338

Milner B (1968) Visual recognition and recall after right temporal-lobe excisions in man. Neuropsychologia 6:191-210

Moore BCJ, Rosen SM (1979) Tune recognition with reduced pitch and interval information. Q J Exp Psychol 31:229-240

Munson WA (1947) The growth of auditory sensitivity . J Acoust Soc Am 19:584

Patterson RD, Peters RW, Milroy R (1983) Threshold duration for melodic pitch. In: Klinke R, Hartmann R (eds) Hearing - physiological basis and psychophysics. Springer, Berlin Heidelberg New York

Phillips WA, Christie DFM (1977) Components of visual memory. Q J Exp Psychol 29:117-133

Robinson DW, Dadson RS (1957) Threshold of hearing and equal loudness relations for pure tones, and the loudness fiction. J Acoust Soc Am 29:1284-1288
Shankweiler D (1966) Effects of temporal-lobe damage on perception of dichotically presented melodies. J Comp Physiol Psychol 62:115-119
Stevens SS, Davis H (1938) Hearing, its psychology and physiology. Wiley, New York
Wechseler D (1964) Die Messung der Intelligenz Erwachsener. Huber, Bern
Wieser HG, Yasargil MG (1982) Selective amygdalohippocampectomy as a surgical treatment of mesiobasal limbic epilepsy. Surg Neuro 17: 445-457.
Winer BJ (1971) Statistical principles in experimental design. McGraw-Hill, New York
Wittlieb-Verpoort E (1989) Attention, memory and temporal estimation in patients with focal brain damage. PhD thesis, University of Zurich
Wittlieb-Verpoort E, Perret E (1987) Cognitive strategies for monitoring auditory signals. Percept Motor Skills 65:847-853

Music Processing After Brain Damage: The Case of Rhythm Without Melody

Isabelle Peretz and Sylvie Hébert

Brain injury can drastically impair musical activities while sparing intellectual and linguistic abilities (see Marin 1982, and Basso 1993, for reviews). Conversely, musicianship can be maintained following brain insult that results in severe global cognitive decline (Beatty, Zavadil, Bailly and al. 1988). These facts suggest that musical abilities are specific functions that are subserved by a neural architecture dedicated to music processing (Peretz and Morais 1993; Sergent et al. 1992). Music is, however, not a monolithic entity. Examination of amusic patients leads to the fascinating realization that music cognition is not affected in its entirety, but in particular abilities only. The selective nature of some of these disorders will be the focus of the present paper. The central idea is that these deficits provide a rich source of information for the understanding of *normal* music cognition.

Music Processing after Brain Damage

At the same time as language impairments subsequent to brain damage were discovered to be independent of impairments in intellectual faculties, disorders of musical abilities were also reported (Bouillaud 1865, cited in Dorgeuille 1966). Yet, in these early days, music disturbances did not spur an intrinsic interest, in that they remained related to, and sometimes integrated with, the concepts guiding classification of language disturbances. At the beginning of this century, most observations still attempted to pinpoint the neuroanatomical localization of musical functions without a real concern for characterizing these functions (see Peretz 1985, for a review). The few exceptions (e.g., Botez and Wertheim 1959; Wertheim and Botez 1961) still suffered from a lack of theoretical development and methodological tools for studying music, and thus were limited, for the most part, to descriptions of an anecdotal nature.

This situation has drastically changed in recent years, both with the rise of interest for the musical domain in normal experimental psychology (see Sloboda 1985; Dowling and Harwood 1986, for textbooks) and the revival of neuropsychological studies as a mean to understand normal cognition (Ellis and Young 1988; Shallice 1988). The rationale behind this endeavor is that any type of dysfunction cannot occur as the result of brain damage; the nature of the disorder is constrained by the premorbid brain. In other words, the observed deficit is constrained by the organization of the normal cognitive system. Following this approach, neuropsychological data can only be exploited if theories of normal functioning exist. Without an implicit or explicit model of how a normal system operates, it is difficult to conceptualize how it may be disrupted. The drawback is that for certain functions, such as music, general models are still lacking. However, normal experimental psychology has provided a series of concepts and tools that can serve as fruitful guidelines for investigating music disturbances. In turn, neuropsychological data can contribute to the building of such models, as we will try to demonstrate in this paper.

As mentioned at the outset, brain damage may offer unparalleled opportunities to observe functional selectivity and hence evidence for componential organization of musical abilities. To illustrate these phenomena, the cases of two brain-damaged patients displaying reverse patterns of breakdown will be briefly described. The first case is that of Shebalin, the Russian composer who, following damage to the left hemisphere, was completely aphasic for the rest of his life; he could neither understand nor speak intelligibly. He nevertheless continued to compose, notably completing his fifth symphony, which Shostakovitch considered to be one of his most brilliant and innovative works (Luria et al. 1965). Shebalin therefore displayed severe language deficits yet retained his musical skills to a remarkable degree. This dissociation cannot simply be explained by the fact that Shebalin was "abnormally musical" from the start. Indeed, the inverse dissociation has recently been observed in a person devoid of any special talent, be it linguistic or musical. This person, C.N., is a young nurse who never received any musical education. As a consequence of successive damage to both temporal lobes she was unable to pick out the simplest over learned tune. She no longer recognized excerpts from her own record collection, unless the music was accompanied by words, nor could she sing the children's songs that she used to enjoy singing with her young son. Yet C.N. did not encounter the least difficulty in understanding and communicating verbally. This retention of perfect language abilities has enabled her to resume her professional career (Peretz et al., submitted).

The straightforward implication of this brief description is that language and music do not necessarily share the same processing mechanisms. Rather they enjoy functional autonomy. This conclusion may appear trivial to most of us, who assume that music and language must obviously be distinct functions. Accordingly, the interest of cases such as those of Shebalin and C.N. is not simply to offer behavioral evidence for such a claim (which is, however, difficult to prove in standard psychology), but to open the way for determining the functional locus at which music and language dissociate. For example, if it were shown that C.N.'s amusia resulted from a perturbation of the mechanisms involved in 'auditory scene analysis' as described by Bregman (1990), such an observation would be far from trivial. It would suggest that language and music possess their own scene analysis mechanisms (as argued, in fact, by Liberman and Mattingly 1989, for speech). Consequently, the interest of a given dissociation depends on the theoretical context adopted. Moreover, dealing with two global functions - such as language and music - is not very informative in itself since the distinction is too broad to be instructive. As will be seen, music is not an indivisible skill but is subserved by numerous components that may be dissociated or isolated. This will be illustrated with C.N., who happened to be disturbed mainly on the melodic dimension and not on the rhythmic dimension.

The Processing of Rhythm and Melody: Some Background

Normal Studies

The melodic dimension - arising from pitch variations - and the rhythmic dimension - due to temporal variations - have long been treated independently, both in theory (for example, Deutsch and Feroe 1981, for pitch organization models, and Longuet-Higgins and Lee 1982, for temporal organization models) and in practice (see any textbook on music cognition, such as Dowling and Harwood 1986). Recently, this tradition has been seriously called into question by Jones, Boltz and their collaborators (Jones et al. 1982; Boltz and Jones 1986; Jones 1987; Boltz 1989a). They argue, quite persuasively, that melody and rhythm are not psychologically independent, but are integrated in music perception and memory; i.e. listeners treat melody and rhythm as a unified dimension in musical sequences.

The few studies which have considered both melody and rhythm in combination have yielded mixed results with respect to this issue of integration. In support of the integration position are findings that rhythmic patterns which conflict with melodic structure yield a much poorer performance than patterns in which rhythm and melody coincide (e.g., Deutsch 1980). Other studies have obtained results that are more compatible with the independent view, in that melodic and rhythmic structures were found to exert distinct influences (e.g., Monahan and Carterette 1985). Task factors cannot easily account for the discrepancy between the two sets of studies, because evidence for integration (Boltz 1989b) has been obtained with the same task that was used to illustrate separation (Palmer and Krumhansl 1987).

One possibility that would allow reconciliation of both sets of results within a single framework is to conceive melody and rhythm as separate perceptual components that are later integrated to produce a unified experience of the musical segment. This view requires that the observation of independent influences or interactive effects is dependent upon the processing stage consulted by the listener, or tapped by the task requirements. The problem with such an account is that, presently, we do not know at what stage in the processing of musical information the percept is integrated. Hence, perceptual independence might prove to be difficult to verify, at least in the normal brain.

Neuropsychological Studies

Studies of brain-damaged patients can help to clarify this situation by way of dissociation. The dissociations are of two basic types: selective loss and selective sparing. If a patient can no longer accomplish operation X (selective loss) but is able to perform operation Y (selective sparing) then the dissociation of X from Y suggests that the two operations are performed by different mechanisms. The inference is all the stronger if another patient is found to display the inverse dissociation, that is to say, the loss of Y and the sparing of X. These two patients then become an instance of a double dissociation. The cases of Shebalin and C.N., mentioned earlier, fulfill these conditions for music and language.

The logic of the argument is quite simple, the validity of the inferences on normal functioning is not. First of all, the dissociations concern the performance on tasks and not the operations under study. The problem is that any given task implies the involvement of multiple operations. Consequently, a dissociation between two tasks is only interesting to the extent that it is not self-evident and enables inferences to be made about a particular component of a mental operation. Second, studying the performance of brain-damaged patients in order to make inferences about normal functioning implies the acceptance of a series of postulates (Caramazza 1984) that may turn out to be too rigid. For instance, selective loss is assumed to result from elimination of an operation and not from reorganization occurring as an adaptation to brain damage. If some reorganization does occur, then the data collected with brain-damaged patients may only indirectly relate to the way a normal system operates. In any event, dissociation remains the preferred method for studying brain-damaged patients since it often occurs in an extremely robust and clearly quantitative fashion.

We may therefore apply this dissociative logic to the case of integration between melody and rhythm. In this research, we relied on simple characterizations of melodic and rhythmic structures, as arising from sequential variations along the relevant dimension, because our purpose was mainly to explore their dissociability after brain damage. With this in mind, we found various elements in the neuropsychological literature that were more compatible with the notion that melody and rhythm are separate rather than integrated.

The first suggestions of dissociations between the processing of melody and rhythm arose from expressive musical behaviors. In singing, rhythmic variations can be preserved while melodic variations are impaired (Mann 1898; Jossman, 1926, reported in Dorgeuille 1966); conversely, melodic variations can be spared while rhythmic variations are defective (Mavlov 1980; Brust 1980). Similarly, a selective impairment of rhythm (Dorgeuille 1966; Brust 1980), or of melody (Dorgeuille 1966; Assal 1973) may occur in reading music. Thus, there is evidence that melody and rhythm are dissociable, although the indications are indirect because they address issues somewhat remote from music perception.

There is, however, more recent evidence that this double dissociation extends to perception. The first indications were provided anecdotally. Sacks (1985) reports the case of O'M, who "complained that recently the hymns in the chapel seemed more and more alike so that she could scarcely distinguish them by tone or tune, but had to rely on the words, or the rhythm" (pp. 135-136). Similarly, Mavlov (1980) reports the case of a musician who developed a severe difficulty for processing rhythms but who could still process pitch sequences adequately. Conversely, Fries and Swihart (1990) describe the case of an amateur musician in whom rhythm discrimination was intact and melody processing impaired.

In two recent studies we have replicated the phenomenon, but in a more systematic fashion. The task involved in both studies was of the "same-different" classification type. Subjects listened to two short passages and had to judge whether they were identical or different. Unfamiliar musical material was used. In the first study (Peretz 1990), the musical stimuli were altered along one dimension - melody or rhythm - in a way relevant for discrimination. Comparison of the results of unilaterally brain-damaged patients with those of neurologically healthy subjects (matched in terms of age and education) revealed dissociations that are considered robust. Some patients scored well below the normal range (established by control subjects) in the deficient dimension, but within the normal range in the unaffected dimension. Two patients were found to be impaired when the discrimination cue was melodic but not rhythmic and two other patients were discovered to display the reverse pattern. Consequently, brain damage can produce a selective loss for either melody alone or rhythm alone. The evidence strongly supports the conclusion that these two dimensions involve the operation of perceptual systems that are at least partially separable.

Rhythm Without Melody

In the second study (Peretz and Kolinsky 1993), a further case, C.N., was investigated. In C.N., we were able to show that such dissociation could be still observed in situations that promote integration of the melodic and rhythmic dimensions in a normal brain. The patient C.N., briefly described earlier, is a young nurse who had undergone successive brain surgeries to have aneurysms clipped on the middle cerebral arteries on each side. After the second operation, C.N. presented with persistent amusia but no other neuropsychological deficit. The computed tomograph (CT) scan examination revealed sequelar lesions in both temporal lobes.

The ability of C.N. to process the melodic and rhythmic dimensions of musical sequences had been extensively examined in a series of experiments in which the relevant dimensions were independently manipulated. As an example of a task that examined the melodic dimension, two melodies that had to be judged as "same or different" were presented: when different, a pitch change was inserted in the second melody. This pitch change either respected the melodic contour, in which the direction in pitch was the same as that of the standard but the specific pitch was different, or violated the contour, in which the direction in pitch and hence the specific pitch differed from the standard. The "contour-respected" sequence is more difficult to discriminate than the "contour-violated," because there is only one cue available (the specific interval), rather than two (the specific interval and the pitch direction). C.N. was found to be completely unable to employ any of these melodic cues to discriminate the musical sequences. Overall, her performance was not significantly different from chance level in any of the tasks in which the perception of pitch variations was critical. On the contrary, in the tasks where rhythmic variations were the only cues available, her performance was well above chance level and in the normal range of the control subjects.

Therefore, C.N. exhibited the dissociation between rhythm and melody in a clear-cut fashion. However, as discussed earlier, it is likely that rhythm and melody are integrated in a normal brain. Because C.N. had a severe disturbance in the processing of the melodic dimension, integration effects cannot occur, producing a "disconnection-like" behavior. To document this phenomenon, we needed to explore how rhythm and melody were jointly processed by C.N., in a situation in which normal listeners show integration effects. As stated earlier, such situations are not readily derived from the literature on normal subjects because opposite results have been observed with identical task parameters. Therefore, keeping with a "same-different" classification task and with the material used previously, we created a Stroop-like situation.

Typically, the Stroop technique consists of demonstration that variations along an irrelevant dimension have deleterious effects on the responses to the relevant dimension. This interpretation goes back to the original findings of Stroop (1935) who showed that naming colors is dramatically impaired if the colored stimulus happens to be a word that names a different color from the actual color (e.g., the word "green" printed in red). The term "Stroop interference" has since been generically applied to any drop in performance attributable to the integration of an irrelevant dimension which is incongruent with, and calls for the opposite response, to the relevant dimension. Indeed, it is unlikely that reading words and perceiving colors are the product of a single system, hence the attribution of the Stroop effect to the integration of distinct and incongruent responses.

Stroop logic can also be applied to the processing of melody and rhythm in a "same-different" classification task. Stroop interference may occur when responses "same" (S) or "different" (D) conflict with the dimension to which subjects are required to attend and the dimension that must be ignored. In principle, such a task can be performed following either rhythm or melody as the relevant dimension for comparison. In the present case, however, the relevant dimension had to be rhythm, in order to allow C.N. to perform reliably. To ensure valid comparisons, control subjects were also required to ignore melodic information and to classify sequences according to rhythm alone.

A scheme of the paradigm is depicted in Fig. 1. An example of target melody is presented which, when followed by one of the four possible comparisons, created one particular trial representative of each of the four experimental conditions (SS, DS, SD and DD). Because Stroop interference is expected to arise from the failure to ignore melodic variations, responses should be more accurate when they are

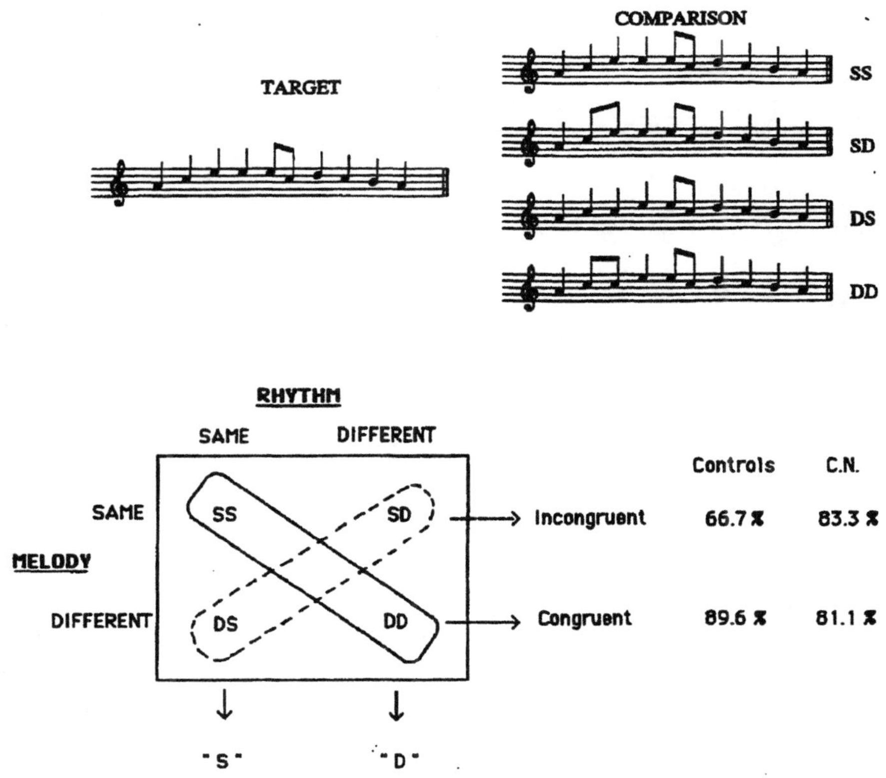

Fig. 1. Example of trials used in the "same - different" task in order to create Stroop-like interference between melody and rhythm. On *top* the target melody and four comparative versions are represented. The conditions yielding the Stroop effect along with the results of the normal matched control subjects and those of C.N. are seen on the *bottom*. *S*, same; *D*, different

congruent (i.e. SS and DD) than when they are incongruent (SD and DS), as illustrated in Fig. 1.

This incongruity interference effect was found to be quite large in the four normal controls (see Fig. 1). Each of the control subjects exhibited this effect. This indicates that melody and rhythm are integrated in an intact brain. In contrast, C.N. did not show evidence of a conflict: she was as accurate on the incongruent as on the congruent trials (see Fig. 1).

The conclusion to be drawn from these results is that, in a *normal* brain, the paradigm was remarkably efficient in producing behavioral evidence of integration between rhythm and melody. Normal subjects exhibited response patterns that were affected by the irrelevant melodic dimension, although decisional independence was promoted by explicitly instructing them to concentrate only on rhythm. The fact that a large incongruity effect was found indicates that rhythm and melody are not entirely separable, but rather are integrated at some level in the processing of

Happy Birthday

Rhythmic condition

Melodic condition

Rhythm and Melody

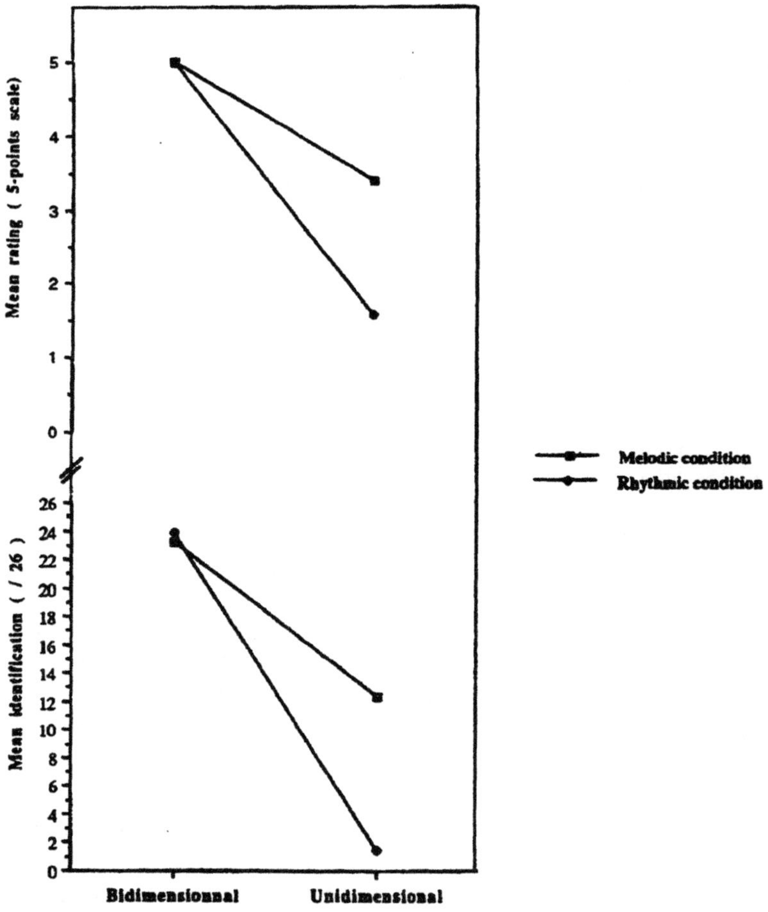

Fig. 2. "Happy Birthday" as presented in the rhythmic and melodic condition to normal subjects for identification and recognition. The results are summarized below.

musical information. C.N. did not exhibit any of the manifestations of integrated behavior observed in normal subjects. Moreover, her performance remained relatively accurate and constant across experiments, thus indicating little, if any, sensitivity to the presence of melodic variation per se. For her, melody and rhythm are entirely separable.

Thus, the results obtained in neuropsychology strongly support the notion that, at some level in the processing of musical information, melodic structure and temporal structure are treated independently. While the notion that rhythm and melody are integral components that are the product of an indivisible working system does not appear to be likely, melody and rhythm are not completely independent dimensions either. Integration was easily elicited in normal subjects. Therefore, a complete independent model for the processing of rhythm and melody is probably no more tenable than is a single-system model.

Finally, C.N.'s results (along with those of another similar case, G.L., described in Peretz et al., submitted; Peretz 1993), raises a new question. C.N. was severely agnosic for music, and she could hardly pick out highly familiar melodies such as "Happy Birthday" illustrated in Fig. 2 (Peretz et al., submitted). Yet, C.N. was not arrhythmic. In principle, she should have used the rhythmic route (i.e., the rhythmic pattern) for recognizing the tunes, but she did not. This finding raises the possibility that melodic information is mostly responsible for tune recognition.

The notion that melody is more determinant than rhythm for tune recognition is plausible in our musical culture. The Western system is indeed rich in the elaboration of pitch variations but relatively poor, or stereotyped, in the rhythmic dimension (while the reverse may occur in African music, for example). Consequently, pitch cues may well be more distinctive than rhythmic cues for accessing or selecting a tune representation in memory. In support of this idea is the often-cited study of White (1960) who found that normal listeners do indeed correctly identify 88% of the tunes when presented with melodic variations in isolation (e.g. the melodic condition of "Happy Birthday" in Fig. 2), whereas they achieve only 33% on the basis of rhythm (e.g., see the rhythmic condition in Fig. 2). The finding that listeners nonetheless identified a third of the tunes on the basis of rhythm may be attributed to guessing. In White's study, only ten tunes were presented and their titles were proposed in the form of a multiple-choice to the subject.

We assessed this interpretation in our laboratory. Normal subjects were randomly assigned to either a rhythmic condition or a melodic condition in which the rhythmic patterns or melodic variations of 26 highly familiar tunes (such as "Happy Birthday") were presented. After the test, subjects' familiarity with the tunes was checked with the "complete" version of the tunes (i.e., with tunes varying in both rhythm and melody). The task of the subjects was first to name the tune and then to rate it on a 5-point familiarity scale, with 5 indicating highly familiar. As can be seen in Fig. 2, subjects succeeded in identifying only one tune, on average (4%), from their rhythmic pattern, whereas they achieved identification of half of them (45%) from hearing the melodic variations alone. The same pattern of results emerged on the familiarity ratings.

Therefore, and as pointed out by C.N.'s data, access to tune representations appears to be largely governed by melodic cues. Rhythmic information is of little help. Given the critical role attributed to melodic information in normal access to tune representations, any degradation of this particular type of musical information should thus result in tune recognition failure. This was indeed the case of C.N. (as well as of G.L. - the other amelodic case mentioned previously).

Conclusion

Studying musical processing in brain-damaged patients provides important insights about the way music is interpreted in a normal brain. In particular, as illustrated in the present paper, neuropsychology provides clear support for the idea that music processing is too coarse a level of analysis; the function decomposes into several separate subabilities, which in turn are carried out by separate subsystems. We have shown here that neuropsychological dissociations can provide a means of teasing apart normally interacting systems. This logic may be successfully applied to the processing of melodic and temporal patterns in music discrimination. In addition, there is reason to reject the widely held notion that familiar tunes can be identified from their characteristic rhythmic pattern alone (Lenneberg 1967; Sturges and Martin 1974; White 1960). These findings illustrate that neuropsychological data can provide important sources of information in building a theory of music cognition.

Acknowledgments.This paper was written while the authors were both supported by the Natural Sciences and Engineering Research Council of Canada.

References

Assal G (1973) Aphasie de Wernicke chez un pianiste. Rev Neurol (Paris) 29:251-255

Basso A (1993) Amusia. In: Spinnler H Boller F (eds) Handbook of neuropsychology, vol 8. Elsevier, Amsterdam (in press)

Beatty W, Zavadil K, Bailly R, Rixen G, Zavadil L, Farnham N, Fisher L (1988) Preserved musical skill in a severely demented patient. Int J Clin Neuropsychol 10:158-164

Boltz M (1989a) Perceiving the end: Effects of tonal relationships on melodic completion. J Exp Psychol 15:749-761

Boltz M (1989b) Rhythm and "good endings": Effects of temporal structure on tonality judgments. Percept Psychophys 46: 9-17

Boltz M, Jones MR (1986) Does rule recursion make melodies easier to reproduce? If not, what does? Cogn Psychol 18:389-431

Botez MI, Wertheim N (1959) Expressive aphasia and amusia following right frontal lesion in a right-handed man. Brain 82:186-202

Bregman AS (1990) Auditory scene analysis: The perceptual organization of sound. MIT Press, Cambridge /MA

Brust J (1980) Music and language: musical alexia and agraphia. Brain 103:367-392

Caramazza A (1984) The logic of neuropsychological research and the problem of patient classification in aphasia. Brain Lang 21:9-20

Deutsch D (1980) The processing of structured and unstructured tonal sequences. Percept Psychophys 28:381-389

Deutsch D, Feroe J (1981) The internal representation of pitch sequences in tonal music. Psychol Rev 88:503-522

Dorgeuille C (1966) Introduction à l'étude des amusies. Doctoral dissertation, University Paris

Dowling W, Harwood D (1986) Music cognition. Academic, New York

Ellis AW, Young AW (1988) Human cognitive neuropsychology. Erlbaum, Hove

Fries W, Swihart AA (1990) Disturbance of rhythm sense following right hemisphere damage. Neuropsychologia 28:1317-1323

Jones MR (1987) Dynamic pattern structure in music: recent theory and research. Percept Psychophys 41: 621-634

Jones MR, Boltz M, Kidd G (1982) Controlled attending as a function of melodic and temporal context. Percept Psychophys 32:221-218

Lenneberg EH (1967) Biological foundations of language. Wiley, New York

Liberman A, Mattingly I (1989) A specialization for speech perception. Science 243:489-494

Longuet-Higgins H, Lee CS (1982) The perception of musical rhythms. Perception 11:115-128

Luria A, Tsvetkova L, Futer J (1965) Aphasia in a composer. J of Neurol Sci 2:288-292

Marin O (1982) Neurological aspects of music perception and performance. In Deutsch D (ed) The psychology of music. Academic, New York

Mavlov L (1980) Amusia due to rhythm agnosia in a musician with left hemisphere damage: a non auditory supramodal defect. Cortex 16:321-338

Monahan C, Carterette E (1985) Pitch and duration as determinants of musical space. Music Percept 3:1-32

Palmer C, Krumhansl C (1987) Independent temporal and pitch structures in determination of musical phrases. J Exp Psychol [Hum Percept and Performance] 13:116-126

Peretz I (1985) Asymétrie hémisphérique dans les amusies. Rev Neurolog (Paris) 141:169-183

Peretz I (1990) Processing of local and global musical information by unilateral brain-damaged patients. Brain 113:1185-1205

Peretz I (1993) Auditory agnosia: A functional analysis. in McAdams S, Bigand E (eds). Cognitive aspects of human audition. Oxford University Press, Oxford

Peretz I, Kolinsky R (1993) Boundaries of separability between melody and rhythm in music discrimination: a neuropsychological perspective. Q J Exp Psychol, 46A, 301-325

Peretz I, Morais J (1993) Specificity for music. In: Spinnler H, Boller F (eds) Handbook of neuropsychology, vol 8. Elsevier, Amsterdam (in press)

Sacks OW (1985) The man who mistook his wife for a hat and other clinical tales. Summit, New York

Sergent J, Zuck E, Terriah S, MacDonald B. (1992) Distributed neural network underlying musical sight reading and keyboard performance. Science 257, 106-109.

Shallice T (1988) From neuropsychology to mental structure. Cambridge University Press, Cambridge

Sloboda J (1985) The musical mind: The cognitive psychology of music. Oxford University Press, London

Stroop J (1935) Studies of interference in serial verbal reactions. J Exp Psychol 18:643-662

Sturges PT, Martin JG (1974) Rhythmic structure in auritory temporal pattern perception and immediate memory. J Exp Psychol 102:377-383

Wertheim N, Botez M (1961) Receptive amusia: A clinical analysis. Brain 84:19-30

White B (1960) Recognition of distorted melodies. Am J of Psychol 73:100-107

Immune Parameters, Musical Abilities, and Anomalous Dominance

Marianne Hassler

Summary

In young adult musicians and nonmusicans of both sexes we investigated the relationship of musical talent to immune parameters and indicators of anomalous dominance. In addition to the behavioral markers asthma/allergies, migraine, and myopia, we measured Immunoglobulin (IgE) and total Ig in blood serum as indicators of immune vulnerability. To test predictions from the Geschwind model of cerebral lateralization (Geschwind and Galaburda 1985), we used a formal path model suggested by McManus and Bryden (1991). In males, the model was in part supported by the data. Musical talent was significantly correlated with atypical language dominance and atypical handedness, and atypical handedness was significantly correlated with immune diseases. Correlations between musical talent and immune parameters were in the expected direction, though they did not reach significance in males and females. Missing significant correlations between conditions in females seem to suggest the Geschwind model may be more appropriate for males than for females.

Introduction

In biographies of famous composers we sometimes find comments on immune vulnerability in musicians. Leonard Bernstein, for example, suffered from asthma from childhood on ; and as a young man he was not called up because the disorder was too serious (Bernstein 1982). Alban Berg suffered from exzemas lifelong, and his death was connected with this disease (blood-poisoning after a treatment; Scherliess 1975). Other biographers of musical composers confined themselves to vague remarks on the poor constitution of their subjects, for instance, Chopin (Bourniquel 1959), Mahler (Schreiber 1971), Schubert (Schneider 1958) or, with respect to emotional problems, Schumann (Boucourechliev 1958), to name just some of them. All their biographers would possibly be surprised at the complex model of the biology of cerebral lateralization provided by Geschwind and Galaburda in 1985. This model suggests that the relationship of musical talent to immune vulnerability does not occur by chance.

Instead, a common developmental mechanism underlying talent, immune dieseases, and anomalous brain organization was hypothesized by these scientists. The developmental mechanism was assumed to be related with the gonadal steroid testosterone. Enhanced prenatal testosterone values and/or effects after week 20 of gestation would modify the normal brain anatomy. Testosterone would slow growth within certain areas of the left cerebral hemisphere and, as a compensatory effect, would promote enlargement of other cortical areas, in particular, homologous regions in the right hemisphere. Such a process would result in anomalous brain organization for verbal and nonverbal functions, i.e., left hemisphere dominance for language processing, and right hemisphere dominance for spatial and other nonverbal functioning would be reduced or even turned to the opposite. In addition, non-righthandedness would be likely to occur instead of strong righthandedness.

As an effect of compensatory right hemispheric growth, Geschwind and Galaburda (1985) have argued that giftedness for those talents might develop for which the right hemisphere is particularly important. Such a talent is, among others, musical ability.

As a side effect of slowed growth of the left hemisphere, development of the thymus gland and other structures connected with the immune system would be deficient (Geschwind and Behan 1982), and vulnerability to atopic diseases (the allergies, typically of childhood onset), autoimmune disorders, and other conditions in which pathogenesis immunity plays a major role would occur in the affected subjects.

With respect to musical capacities, the Geschwind model predicts musicians are more likely to have anomalous brain organization for language and nonverbal processing than non-musicians, musicians are more often non-right-handed than non-musicians , and that they suffer more often from immune diseases than non-musicians.

The Geschwind model attempts to explain a vast range of conditions and processes that not had previously been linked, either empirically or theoretically (McManus and Bryden, 1991); it includes talents but also learning disorders and emotional problems, among others. To support their hypothesis, Geschwind and Galaburda have provided a considerable amount of experimental data from various research areas within the realm of neuroscience. I have restricted my references and only offer some examples pertaining to my present topic: To begin with the most surprising aspect of the model, i.e., with immune vulnerability in subjects with anomalous dominance, Geschwind and Galaburda (1985) provided data from animal research showing that the intact left cerebral hemisphere is important for intact immune responses; they also provided evidence that sex hormones influence immune functioning. Recent studies have supported their data: Renoux (1988) found a lateralized influence of the neocortex on immune function in mice. In left-lesioned, but not in right-lesioned mice, spleen and thymus weights were reduced, and number of T cells were decreased to about 50% of that of controls. McCruden and Stimson (1991) have reported damaging effects of prenatal testosterone treatment of pregnant female mice, leading to immune diseases in the offspring.

With regard to handedness, there are experimental studies showing enhanced occurrence of non-right-handedness in musically talented subjects (Byrne 1974; Deutsch 1980; Hassler and Birbaumer 1988).

With respect to functional brain organization for musical material, differences between musicians and nonmusicians have been shown by Bever and Chiarello as early as 1974. Their data clearly showed the notion that there is always an intimate link between music and the right hemisphere to be an oversimplification. On the other hand, single case studies of musical composers who had suffered from damage to the right cerebral hemisphere seem to indicate the intact right hemisphere is, indeed, important to musical creativity. For example, a young male composer (Gardner 1982) retained his musical knowledge after damage to the right hemisphere, and he was able to teach at a conservatory after a time; however, he had lost his interest in creating music himself. Judd's patient (Judd 1988) began to teach music 3 months after a stroke to the right hemisphere, and in a year he took up conducting an orchestra again. But his musical composition was limited, despite efforts to continue it.

Anomalous dominance for verbal material in musicians has been demonstrated with various neuropsychological procedures, i.e., with dichotic listening, tachistoscopic measurements, and EEG (Gordon 1983, for review; Hassler 1990). Deviations from the standard dominance pattern found in musicians were due to enhanced involvement of right hemisphere functions in verbal processing. This outcome would have been expected from the Geschwind and Galaburda (1985) thesis.

The Geschwind model has proved to be influential and stimulating for neuroscientific research. However, testing predictions from the model experimentally is faced with serious problems arising from the very complexity and size of the model, which implicates the risk of not being open to falsification. McManus and Bryden (1991) have suggested formal path models to make predictions testable and falsifiable. There are ethical difficulties in testing the hypothesis, of course, and experimental manipulations of testosterone pre- and perinatally cannot be conducted in humans. Testing the hypothesis in healthy subjects is restricted to assessment of predicted consequences of testosterone effects on the developing brain.

We have tested the consequences of prenatal or perinatal brain development predicted for musicians. Musical talent was expected to be connected with atypical language dominance, atypical handedness, and vulnerability to immune disorders.

ɛcts (45 males, 48 females) were retested at stage 8 of a longitu-
group of 51 (26 males, 25 females) provided blood samples for
ormone as well as neurotransmitter determination (Hassler et al.
ɩup was selected with respect to handedness (15 left-handers and
handers) and with respect to hemispheric lateralization for ver-
original sample ($n = 93$) consisted of 30 musicians who were able
improvise at the first stage of our study (group 1), of 32 musicians
to compose/improvise (group 2), and of 31 nonmusicians (group
19.14 years (SD 1.67) and 18.64 (SD 0.96), males and females,
ubgroup consisted of 21 subjects from group 1, 15 subjects from
ubjects from group 3. Mean age was 19.16 (SD 1.65) and 18.77
nd females, respectively. For a detailed description and selection
r (1992).

ssessed with four questions from Porac et al. (1980):

ich hand would you throw a ball to hit a target?
ich hand do you draw?
ich hand do you use an eraser on paper?
ich hand do you remove a card when dealing?

ith the right hand" was scored 3 points; answering "with either
2 points; answering "with the left hand" was scored 1 point.
14-9 points formed the category "left-handers"; those who attained
classified as "right-handers" (see Hassler and Birbaumer 1988,

verbal material was measured with a dichotic listening task
airs of one-syllable German words presented simultaneously via
three pairs given in 1.5 s (500 s per pair) each participant was
many of the six words as possible (free recall). A laterality index
cording to the formula: REA = (R-L) / (R+L).

s assessed with tests 1-3 from Wing's Standardized Tests of Mu-
(Wing 1968). These tests measure chord analysis, pitch change,

and memory for melodies. Test scores were transformed to z-scores to account for the broad range within the best category "A" and to avoid ceiling effects.

Immune parameters were assessed with a questionnaire. The subjects answered questions about asthma/allergies, migraine, and myopia based on the diagnosis of a physician. The questions were answered "Yes" or "No."

The Physiological markers IgE and total Ig were measured in blood serum in the laboratory of the Department of Diagnostic Endocrinology, University Children's Hospital, Tübingen, using the Mancini technique. Serum IgE apparently reflects an excess of IgE antibody in a pathophysiological sense, because allergic reactions occur in sensitized tissues (Frick 1987).

Procedure

All participants were tested in small groups of 2, 3, or 4, males and females, separately, by same-sex experimenters with a test battery described elsewhere (Hassler et al. 1985). The dichotic listening task was administered individually. Blood was taken on the day of testing or on the day before or after testing between 8 a.m. and 9 a.m. Blood serum was separated using a centrifuge and then frozen to -20°C.

Statistical Analysis

Test scores of the Wing tests 1-3, handedness scores, the laterality index REA and IgE values were correlated using Pearson correlation coefficients. For asthma/ allergies and the composite score, point biserial correlations were calculated. Data of males and females were treated separately.

Results

1. Relationship of Musical Capacities to Anomalous Dominance

- *Males.* Significant negative correlations between the Wing tests and the two indicators of anomalous dominance, i.e., the laterality index and handedness, showed a relationship of musical capacities to anomalous brain organization for language functions and to left-handedness. This outcome would have been expected from the Geschwind model.
- *Females.* In contrast to the outcome in males no significant correlations between musical talent and anomalous dominance emerged in women.

2. Relationship of Musical Talent to Immune Parameters

- *Males* No significant correlations emerged between Wing test scores and either diseases or physiological immune parameters.
- *Females* as in males, no significant connections were found between musical talent and immune parameters in females.

3. Relationship of Anomalous Dominance to Immune Parameters

- *Males.* Left-handedness was significantly correlated with the composite score, i.e., left-handers suffered more often than right-handers from asthma/allergies or migraine or myopia. This result would have been expected from the Geschwind model.
- *Females.* REA and handedness were not significantly related to immune parameters in women.

Table 1. Means and standard deviations (in parentheses) of variables

	Total group	Subgroup
Males [a]		
Wing 1-3	1.44 (1.01)*	1.69 (.74)*
REA	.07 (.16)**	.06 (.20)
HD	10.07 (2.89)	9.31 (3.48)
IgE (IU/ml)		48.27 (50.37)
Igt (mg/dl)		1169.42 (243.64)
Females [b]		
Wing 1-3	1.16 (.95)	1.16 (1.15)
REA	- .05 (.14)	.01 (.13)
HD	10.98 (1.86)	10.40 (2.36)
IgE (IU/ml)		62.13 (74.71)
Igt (mg/dl)		1118.58 (194.50)

Wing 1-3, musical talent; HD, handedness; REA, verbal laterality index: IgE, immunoglobulin E; Igt, immunoglobulin total
Sex differences: *$p < .10$; **$p < .05$

[a] Total group, $n = 45$; Subgroup, $n = 26$.
[b] Total group, $n = 48$; Subgroup, $n = 25$

Table 2. Correlation coefficients of the Wing tests 1-3 (musical ability) and indicators of anomalous dominance and immune

	Wing 1-3	REA	HD	A/A	comp
Males					
REA t	-.27*				
REA s	-.30				
HD t	-.34**	.25			
HD s	-.54**	.27			
A/A s	-.02	.24	-.05		
comp s	.24	-.02	-.33		
IgE s	.09	-.12	.07	.40**	.28*
Igt s	.21	-.23	-.20	.04	.06
Females					
REA t	.02				
REA s	-.05				
HD t	-.11	.02			
HD s	-.14	.10			
A/A s	-.05	-.10	-.14		
comp s	.19	.01	-.07		
IgE s	.30	-.26	.03	.14	-.07
Igt s	-.17	.11	.23	-.31	-.20

REA, verbal laterality index; HD, handedness; A/A, asthma/allergies; comp, composite score (asthma/allergies, migraine, and myopia, combined); IgE, immunoglobulin E; Igt, immunoglobulin total; t, total number; s, subgroup
*$p < .10$; **$p < .05$

To summarize the outcome: While in males the present data provided some support for the Geschwind model - anomalous brain organization for verbal processing and non-right-handedness were significantly related to musical talent, and non-righthandedness was significantly related to some immune parameters, the female data did not. However, there is a positive correlation between musical talent and IgE values (.30) which probably would have been significant with a larger sample in females.

Means and standard deviations of variables were summarized in Table 1, correlation coefficients were summarized in Table 2, males and females separately.

In Fig. 1, the outcome of the present study was put in a formal path model suggested by McManus and Bryden (1991).

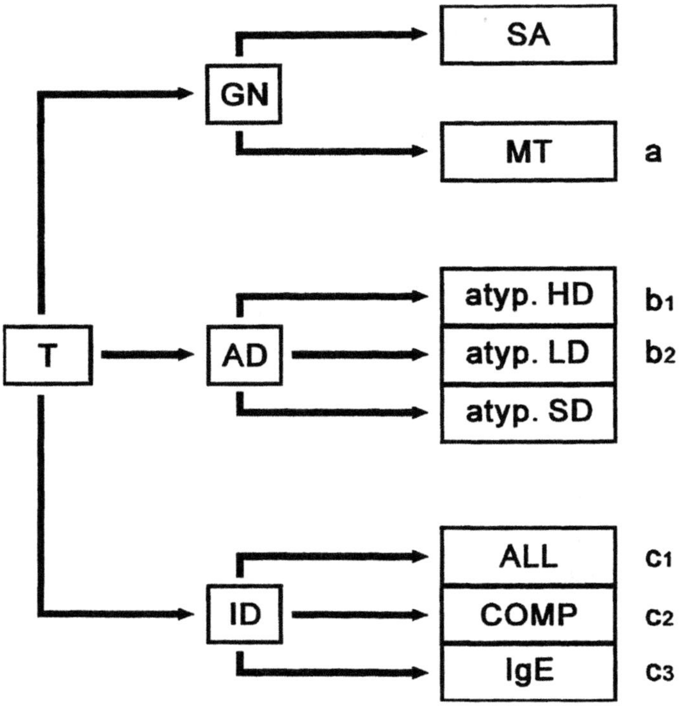

Fig.1. Formal Path Model Following McManus and Bryden (1991). *SA*, spatial ability; *MT*, musical talent; *atyp. HD*, atypical handedness; *atyp. LD*, atypical language dominance; *atyp. SD*, atypical visuospatial dominance; *ALL*, asthma/allergies; *COMP*, composite score (asthma/allergies, migraine and myopia, combined); *IgE*, immunoglobulin E

Predicted relations by the Geschwind model (Geschwind and Galaburda 1985):

GN (giftedness) - AD	(anomalous dominance)	(a - b1, b2)
GN - ID	(immune disorders)	(a - c1, c2, c3)
AD - ID		(b1, b2 - c1, c2, c3)

Obtained relations in the present study; correlations $p < .10$

Males:		Females:	
GN - AD	(a - b1; a - b2)	GN - AD	/
GN - ID	/	GN - ID	/
AD - ID	(b1 - c2)	AD - ID	/

Discussion

The Geschwind model of the biology of cerebral lateralization (Geschwind and Galaburda, 1985) has proposed enhanced prenatal testosterone effects on the developing brain to be the common mechanism underlying musical talent, anomalous brain organization for language functions, non-right-handedness, and immune vulnerability. The model thus provides the basis for a new understanding of data reported separately in biographies of musical composers and in experimental studies.

As in previous research (Benbow 1987; Hassler and Birbaumer 1988; Temple 1990), we were not able to manipulate prenatal testosterone to determine the impact on the developing brain, and we were restricted to testing various associations between conditions postulated to be consequences of prenatal testosterone effects. In men, we found some support for the model. In males, anomalous dominance for language functions and non-right-handedness were significantly related to musical talent, and non-right-handedness was significantly related to immune disorders combined in the composite score. Positive, though not significant, correlations between musical talent and immune parameters emerged as well. Here, the small sample of subjects in the subgroup may have contributed to our failure to find significant results. With a larger sample and, because right hemispherical functions may be especially important for composition (Gardner 1982; Judd 1988), with musical composers as subjects the Geschwind model might be more strongly supported .

The results in females failed to provide evidence supporting the Geschwind thesis. Though correlations between musical talent and immune parameters tended to be in the expected direction and would possibly have been significant with a larger sample, the central thesis of the model was not confirmed. Female musicians, but also female nonmusicians, had anomalous dominance for verbal material, and correlations showed no relationship of musical talent to anomalous brain organization or to non-right-handedness. Indicators of anomalous dominance - atypical language dominance, atypical handedness - were not related to immune parameters, and they were not related to each other.

A possible explanation for sex differences in relationship comes from Witelson's studies of the human corpus callosum (Witelson 1991; Witelson and Goldsmith 1991). These authors measured the corpus callosum in left-handers and right-handers, and the midsagittal area of the corpus callosum was found to be larger in left-handers and ambidextrous individuals than in right-handers (Witelson 1985). Differences were evident only in males, not in females. When correlations were calculated between the respective area and a quantitative handedness score, these correlations were $r = -.67$ ($p<.001$) for males. For females, these correlations were near zero ($r = -.09$; Witelson 1991). Witelson has concluded that this suggests not merely a difference but a dichotomy between the sexes in the neuroanatomical substrate of handedness and, by inference, for aspects of functional asymmetry. The outcome also seems to suggest that prenatal developmental mechanisms leading to anomalous dominance may not be identical in males and females. Our data also seem to indicate this.

To account for the criticism that the very size and complexity of the Geschwind model implicates the risk of not being open to falsification (McManus and Bryden 1991), we used a formal path model suggested by these authors to make predictions testable and falsifiable. The formal path model implicates the notion that there are linear relations between processes and conditions incorporated in the Geschwind model. We should, however, be aware that linear relations between conditions have to be justified in future research. When, for example, testosterone was investigated in relation to human sexual behavior, most studies failed to show significant correlations (see Meyer-Bahlburg 1984, for review). On the other hand, a minimum level of testosterone is absolutely necessary for the existence of sexual motivation and behavior (Nieschlag 1979). Thus, future studies will show whether formal path models (McManus and Bryden 1991) will take advantage over group comparisons (Benbow 1987; Hassler and Birbaumer 1988; Temple 1990) in testing Geschwind's hypothesis or whether new statistical procedures should be introduced.

Acknowledgments. Work supported by a grant from the DFG (German Research Society) Ha 1350. Prof.Dr. Derek Gupta is gratefully acknowledged for cooperation. Dipl.-Psych. Karl Kirschmann is acknowledged for statistics and graphics.

References

Benbow CP (1987) Possible biological correlates of precocious mathematical reasoning ability. Trends Neurosci 10:17-20

Bernstein B (1982) Family matters. Summit , New York

Bever TG, Chiarello RI (1974) Cerebral dominance in musicians and non-musicians. Science 185:537-540

Boucourechliev A (1958) Schumann. Rowohlt, Reinbeck (rororo Bild Monographien)

Bourniquel C (1959) Chopin. Rowohlt, Reinbeck (rororo Bild Monographien)

Byrne B (1974) Handedness and musical ability. Br J Psychol 65:279-281

Deutsch D (1980) Handedness and memory for tonal pitch. In Herron J (ed) Neuropsychology of lefthandedness. Academic , New York, pp 263-272

Frick OL (1987) Immediate hypersensitivity. In: Stites DP, Stobo JD, Wells JV (ed) Basic and clinical immunology, 6th Edn. Appleton and Lange, East Norwalk, pp 197-227.

Gardner H (1982) The music of the hemispheres. Psychol Today June:91-92

Geschwind N, Behan P (1982) Left-handedness: associations with immune disease, migraine, and developmental learning disorder. Proc Natl Acad Sci USA 79:5097-5100

Geschwind N, Galaburda AM (1985) Cerebral lateralization. Biological mechanisms, associations, and pathology: I. A hypothesis and a program for research. Arch Neurol 42:428-459

Gordon HW (1983) Music and the right hemisphere. In Young AW (ed) Functions of the right cerebral hemisphere. Academic, New York, pp 65-86

Hassler M (1990) Functional cerebral asymmetries and cognitive abilities in musicians, painters, and controls. Brain Cogn 13:1-17

Hassler M (1992) Creative musical behavior and sex hormones: musical talent and spatial ability in the two sexes. Psychoneuroendocrinology 17:55-70

Hassler M, Birbaumer N,Teil A (1985) Musical talent and visual-spatial ability: a longitudinal study. Psychol Music 13:99-113

Hassler M, Birbaumer N (1988) Handedness, musical abilities, and dichaptic and dichotic performance in adolescents: a longitudinal study. Dev Neuropsychol 4:129-145

Hassler M, Gupta D, Wollmann H (1992) Testosterone, estradiol, ACTH and musical, spatial and verbal performance. Int J Neurosci 65:45-60
Judd T (1988) The varieties of musical talent. In Obler LK, Fein D (eds) The exceptional brain. Guilford, New York, pp 127-155
McCruden AB, Stimson WH (1991) Sex hormones and immune function. In Ader R, Felten DL, Cohen N (eds) Psychoneuroimmunology, 2nd edn. Academic, New York
McManus IC, Bryden MP (1991) Geschwind's theory of cerebral lateralization: developing a formal causal model. Psychol Bull 110:237-253
Meyer-Bahlburg HFL (1984) Psychoendocrine research on sexual orientation. Current status and future options. In: de Vries GJ, de Bruin JPC, Uylings HBM, Corner MA (eds) Sex differences in the brain. Elsevier, Amsterdam, pp 375-398
Nieschlag E (1979) The endocrine function of the human testis in regard to sexuality. sex, hormones and behaviour. Ciba Found Symp 62:182-208
Porac C, Coren S, Steiger JH, Duncan P (1980) Human laterality: a multidimensional approach. Can J Psychol 34:91-96
Renoux G (1988) The cortex regulates the immune system and the activities of the T-cell specific immunopotentiator. Int J Neurosci 39:177-187
Scherliess V (1975) Alban Berg. Rowohlt, Reinbeck (rororo Bild Monographien)
Schneider M (1958) Schubert. Rowohlt, Reinbeck (rororo Bild Monographien)
Schreiber W (1971) Mahler. Rowohlt, Reinbeck (rororo Bild Monographien)
Temple CM (1990) Academic discipline, handedness, and immune disorders. Neuropsychologia 28:303-308
Wing H (1968) Tests of musical ability and appreciation, 2nd edn. The Br J Psychol Monogr Suppl 27
Witelson SF (1985) The brain connection: the corpus callosum is larger in lefthanders. Science 229:665-668
Witelson SF (1991) Neural sexual mosaicism: sexual differentiation of the human temporo-parietal region for functional asymmetry. Psychoneuroendocrinology 16:131-153
Witelson SF, Goldsmith CH (1991) The relationship of hand preference to anatomy of the corpus callosum in men. Brain Res 545:175-182

III. Music Psychopathology

The Psychopathology of the Sense of Music

REINHARD STEINBERG

Introduction

Like speech, music in its proper sense owns only to men. Like speech comprehension or production, music perception and performance also requires cognitive, affective and motoric competence. In speech as a result of thinking and in musical performance as a result of affective comprehension and motoricity, a weakening is obvious when an individual becomes mentally ill. For understanding the pathophysiologic processes underlying schizophrenia, depression, or mania, all kinds of verbal and nonverbal communications are listed in detail and ordered in international psychopathologic systematizations. Surprisingly music has been left out completely, although its fundamental importance for human psychic equilibrium is generally accepted. Music is even a very apt model for studying cognitive functions, as research in the field of amusia shows. Its lack of semantic meaning, and its physically well-definable and meaningful rhythmic, harmonic, and melodic constituents may even have some advantages over speech in some aspects of cognition research.

The therapeutic musical activity of mentally ill patients can be used as a research tool in studying music cognition. Mental disease is a naturally occurring alteration of cerebral function, which an affected person communicates through many nonverbal behaviors, also when performing music. Since today´s therapeutic efforts often lead to a complete or at least to a partial remission, there are two well-definable states for intraindividual examination in follow-up design. Any instrumental performance reveals motoric skills of any degree, those of poor training as well as those altered by a disease. It also reveals the degree of understanding and ability to reproduce affective content, which emerge in the attempt to follow mainly unwritten rules, such as tempo variation. The ideal typic "Gestalt" of music even allows for an assessment of achievement, which can be defined easier than with verbal utterances.

It has been hypothesized in several studies (Steinberg and Raith 1985a,b; Steinberg et al. 1991, 1992b; Raith et al. 1989) that the manner of playing music changes systematically according to distinct psychiatric syndromes or diagnoses, which would even allow us to speak of music psychopathology. There is also the idea that some functional and morphologic pecularities correlate with those supposed music psychopathologic features, which will enable a better understanding of cognition in patients and controls (Günther et al. 1988, 1991, 1993; Steinberg et al. 1989, 1992a,b).

Method

Several studys were conducted in over 200 mental patients of various instrumental aptitude and 100 matched controls, which answer some questions and obviously open up many more. Despite a generally severe psychopathologic alteration at admission almost all patients willingly allowed their instrumental performances to be recorded. This was done, in the music room of Munich university hospital. Since the first recording were generally made shortly after admission before the patient had received medication, a significant intervening effect of psychotropic drugs could be excluded. A second recording was made of most patients after about 6 weeks, when symptoms had improved considerably or even disappeared. With different rating scales the psychopathologic state was ascertained by a psychiatrist not attending the performance. Also the rating teams who analyzed the recordings were ignorant of the patient´s psychopathology and instrumental aptitude (Steinberg and Raith 1985a,b; Steinberg et al. 1985).

Results

Depression generally slows down many activities. In western music depressed affect is generally associated with a slow tempo, e.g., a funeral march with the tempo "grave," whereas a fast musical tempo for instance is indicated "allegro," which literally means cheerful. Therefore, the assumption that a depressed mood would slow down musical performance seems easily understood.

In order to define this assumed effect of depression on musical motoricity, the subjects´ recording were analyzed with regard to musical tempo. First, a Maelzel´s metronome was brought into line with the performance tempo, giving its tempo in beats per minute (MM). Since in exaggerated rubato and in a very irregular performance an overall tempo is more easily defined by the fraction between the beats of a piece and its duration in minutes, the latter was measured also in a second step using a stop watch. The fraction between these two tempi results in a stability quotient (SQ), which usually amounts to 1. Shortening of pauses or a continuous accelerando would lower the SQ, and hesitations or a global deceleration would augment it.

It was not really surprising that no general correlation was found between performance tempo and depressive syndromes, since all common instruments, pieces in various musical styles and with different tempo indications and, what turned out to be of considerable importance, people with all level of achievement, whether professional or very amateurish, were contained in the sample. The hypothesized correlation was found by chance, using an easy folk song which all subjects played by heart (Steinberg and Raith 1985a). Also in further examinations under specific

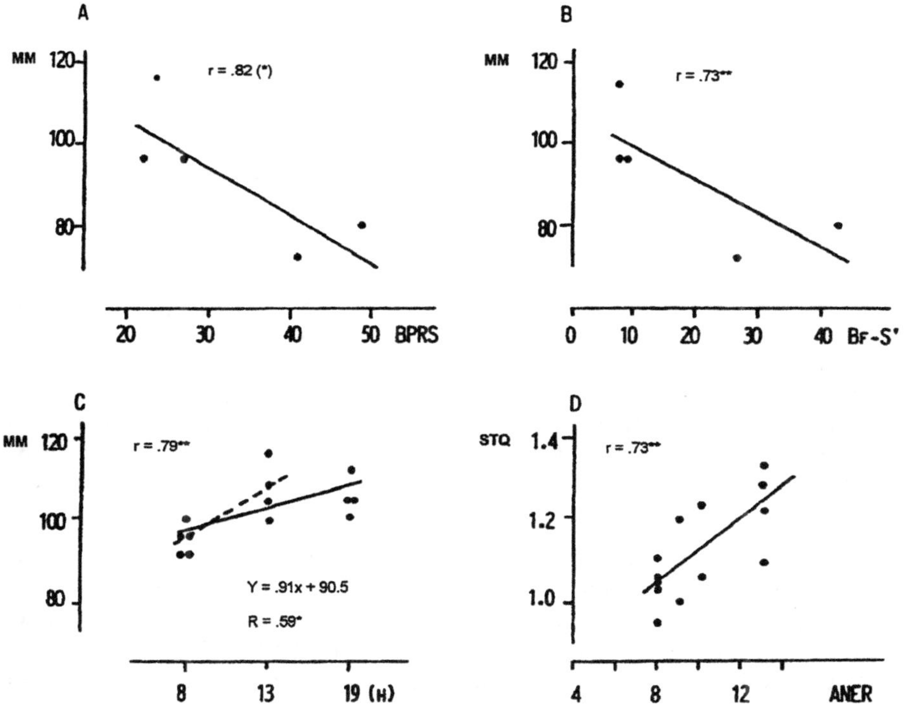

Fig. 1. a Negative correlation between the tempo of the folk song "Hänschen klein," played five times by a 36-year-old violin teacher, and psychopathology assessed by means of the BPRS (Overall and Gorham 1962). **b** Negative correlation between self estimation, assessed by means of Bf-S′ (von Zerssen 1979), and tempo of the folk song. Same patient as in **a**. **c** Correlation between tempo of the folk song and diurnal mood variation; a 29-year-old librarian playing recorder. **d** Correlation between stability quotient (STQ) and the factor ANER from BPRS, same patient as in **c**. See text. (*) $p<0.1$; * $p<0.05$; ** $p<0.01$; *** $p<0.001$

experimental conditions it was proved with composed music (Steinberg et al. 1992b). From the start of the study all patients and controls were asked to perform right at the beginning and then at the end of a recording session the very simple German folk song "Hänschen klein," which all subjects knowed by heart. Figure 1 shows the correlation between performance tempo and the objective and subjective psychopathologic state of a 36-year-old violin teacher during the course of inpatient treatment for severe depression. She played the folk song obviously faster when in remission, which was attained after 6 weeks. Another leading symptom of the melancholic type of major depression, regular, diurnal variations in mood and psychomotor tempo, were apparent in a 29-year-old librarian. In Fig. 1c a significant slowing down of the rendition of the folk song played on the record at 8 a.m. is obvious. This is due to the patient′s global morning anergia, which by noon already had ceased to a considerable extent. Her playing in the morning was characterized

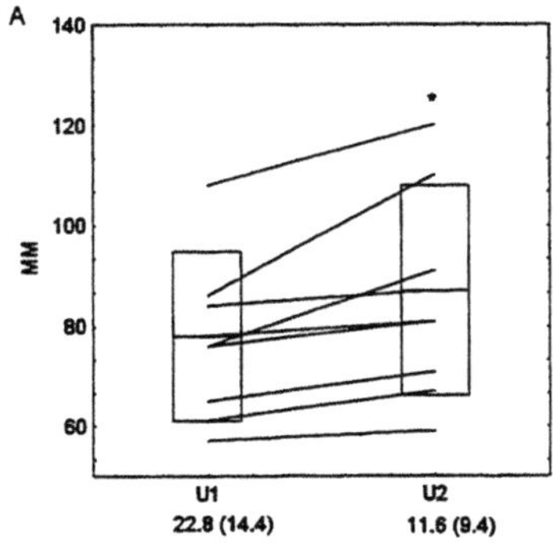

Fig. 2. a Individual and mean tempo of the folk song dependent on diurnal variations. U_1 recorded before 9 a.m., U_2 after 4 p.m. the same day.

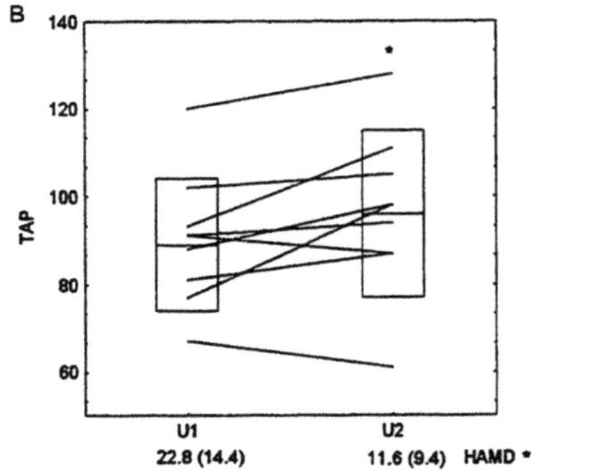

Fig. 2. b Number of taps made with a pencil in maximum speed within 10 s, extrapolated to 1 min. HAMD (Hamilton 1960). For *p* values see Fig. 1

by long phraseologic pauses, which led to a considerable increase in SQ corresponding to the predominant anergic symptom. The SQ had been rated by the patient´s physician independently just before the recording session. With the obvious reduction in anergia quite normal performances were attained (Fig. 1d)

With this folk song very peculiar musical phenomena resulted in general. As has already been published (Steinberg and Raith 1985a), 21 controls played "Hänschen klein" at completely different tempi, between MM 46 and MM 162 with the median at about MM 85. But surprisingly, this individual tempo seems very stable, since after 30 min, 12 h and even 6 weeks the same individual tempo is chosen, representing a stable inner representation of this song. In endogenously depressed patients, the tempo of the children´s song dropped significantly, by about 10 MM on average, in the depressed state, whereas in remission the average tempo played was quite normal. Also, diurnal mood variations were reflected in the significantly faster tempo in the remitted state in the evening as compared to the patient´s deep depression in the morning as indicated by the Hamilton Depression Score (HAMD) (Fig. 2a). This is in parallel with the general motoric inhibition, since tapping at maximum speed also follows the same diurnal patterns (Fig. 2b). The close time frame of the investigation of diurnal variations also revealed the involvement of the tempo of composed music, as is seen in Fig. 3a. In the morning during deep depression the average tempo was significantly lower, by about MM 10, than the performances of the same individually chosen pieces in the evening. The controls did not show this tempo phenomenon (Fig. 3b), nor did any other patient group - not even the neurotic-depressed patients. This obviously different psychomotoric behavior may be an argument for at least two different forms of depression (Steinberg et al. 1985). As is shown in Fig. 4, this phenomenon is confined to deep depression, since a partial remission after about 2 weeks at U_M (recording after 2 weeks) results already in the average tempo of U_2 (recording after 6 weeks) ratings. The professional musicians in the sample did not influence systematically the tempo

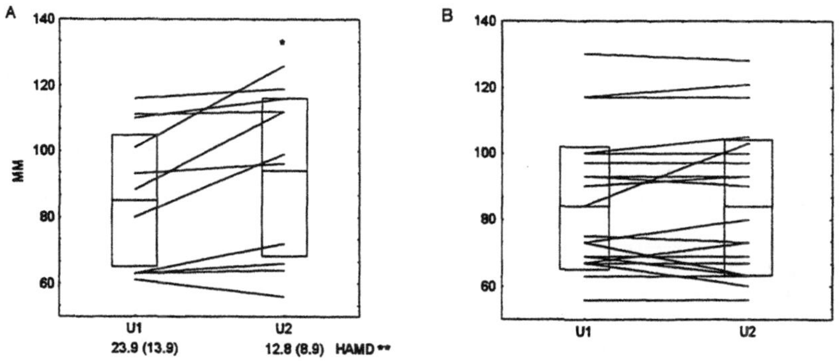

Fig. 3. a,b Individual and mean tempo literate music pieces were played by 11 endogeneously deepressed patients (HAMD; mean SD)(**a**) and 21 controls at U_1 and U_2 (**b**) For *p* values see Fig.

of the folk song: a good instrumentalist did not play any faster than one who was not as good. The tempo of the folk song seems to follow a Gaussian distribution with a mean tempo of around MM 85 (MM 46 - MM 168).

But how can musical achievement be defined and quantified with such a sample of average aptitude, almost all persons playing different pieces in all different musical styles, different requirements to technical performance, and even different instruments. For reasons of time a restriction to good or even professional musicians did not make sense. Over and above that an average sample also had the advantage of being representative of musical behavior in the general population. Musical aptitude of the participants was defined on a seven-point scale by taking into account the time of instrumental instruction, rehearsing, practice, degree of achievement, musical profession and so on. A professional concert soloist was rated 1, whereas an amateur without any instrumental training and musical education rated 7. A formerly well educated, not professional instrumentalist with some actual experience

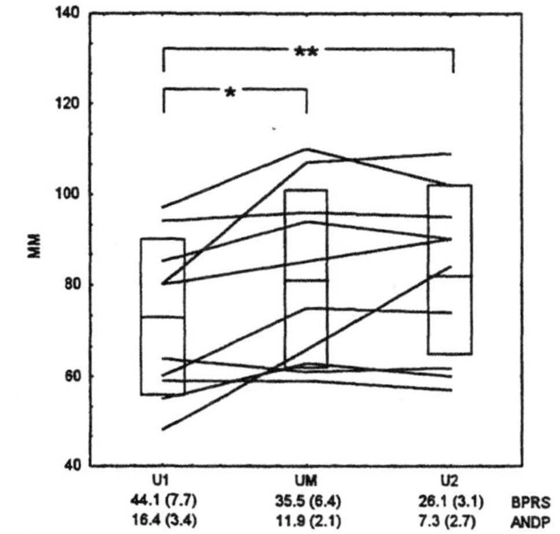

Fig. 4. Individual and mean tempo of the folk song at U_1, U_M and U_2 as played by 10 endogeneously depressed patients. For *p* values see Fig. 1

Table 1. Objective performance characteristics at U1 and U2 for the upper and the lover third of the patients ranked according to delta-BPRS

	Upper third (BPRS) (n = 18) PI = 4.3 (1.6)		Lower third (BPRS) (n = 18) PI = 4.3 (1.2)	
	U_1	U_2	U_1	U_2
BPRS	49.7 (9.5)***	25.0 (3.8)	47.3 (12.5)***	40.4 (8.5)
Formal characteristics				
Repeated beginnings	2.4 (3.0)	1.4 (3.1)	1.4 (4.2)	0.7 (1.9)
Interruptions	2.4 (4.2)	1.6 (3.2)	2.3 (5.7)	1.2 (1.9)
Tempo				
Average tempo	87.5 (26.7)	95.4 (34.0)	101.2 (37.9)	99.2 (40.6)
Opening tempo	90.7 (24.6)	95.4 (28.3)	115.7 (32.4)	110.9 (32.1)
Max. tempo	98.9 (28.6)	108.9 (45.0)	125.6 (39.7)	125.2 (44.2)
Min. tempo	79.6 (22.8)	85.9 (27.0)	84.4 (37.2)	89.2 (33.6)
Acceleration	6.5 (4.4)	5.4 (4.2)	6.8 (2.8)	6.4 (3.5)
Deceleration	9.2 (7.1)	6.6 (3.1)	8.8 (3.8)	7.9 (4.2)
Keeping time				
Rest too short	2.3 (3.6)	2.8 (4.2)	1.3 (2.2)	0.8 (1.2)
Rest too long	0.8 (2.8)	1.0 (1.3)	1.1 (2.8)	1.0 (1.4)
Timing error	0.6 (2.0)	0.2 (0.7)	1.9 (2.3)*	0.7 (1.2)
Left-right shifting	0 (0)	0.2 (0.5)	0.4 (0.9)	0.6 (1.0)
Rhythm				
Hesitation	1.3 (2.3)	1.2 (1.9)	3.1 (4.0)	3.4 (5.7)
Hesit. with error	0.1 (0.5)+	0.8 (1.4)	1.5 (3.3)	1.2 (1.9)
Hesit. with reiteration	0.1 (0.2)	0.3 (1.0)	0.6 (0.9)	0.4 (0.8)
Hesit. + error + reiteration	2.3 (4.1)	0.9 (1.6)	3.3 (4.5)	2.4 (3.9)
Melody				
Sketchy playing	2.4 (3.2)	5.0 (9.7)	4.9 (6.9)*	7.6 (10.8)
Wrong notes	8.5 (12.0)+	3.6 (4.3)	7.4 (10.2)+	3.6 (3.8)
Notes missed out	1.7 (3.1)	2.4 (3.9)	4.4 (8.4)	4.3 (5.0)
Added notes	0.9 (2.1)	0.6 (1.4)	1.3 (1.9)	0.6 (0.9)
Technical aspects				
Irregular accentuation	1.8 (3.3)	1.7 (4.7)	2.8 (3.8)	3.1 (5.1)
Inaccurate playing	7.0 (8.6)*	3.2 (3.9)	2.7 (3.3)	3.6 (4.5)
Technique	5.2 (2.6)+	4.7 (2.5)	5.3 (1.9)	4.9 (2.0)

see legend to Fig. 1 for *p* values.

BPRS, Brief Psychiatric Rating Scale; PI, professionality index

in performance represents the average of 3.5, which was gained in the whole sample and, which is most important for comparison of syndromatic or nosologic subgroups, in all diagnostic and control samples (Steinberg et al. 1985).

But an average aptitude also implies an average performance. This otherwise means the general occurrence of playing mistakes, irregularities and peculiarities of amateur performance to a considerable extent. Therefore playing mistakes and performance irregularities had to be taken into consideration (Steinberg et al. 1992a,b). The hypothesis that an acute state of a mental disease weakens musical expression would find a banal explication if only fewer playing mistakes in the

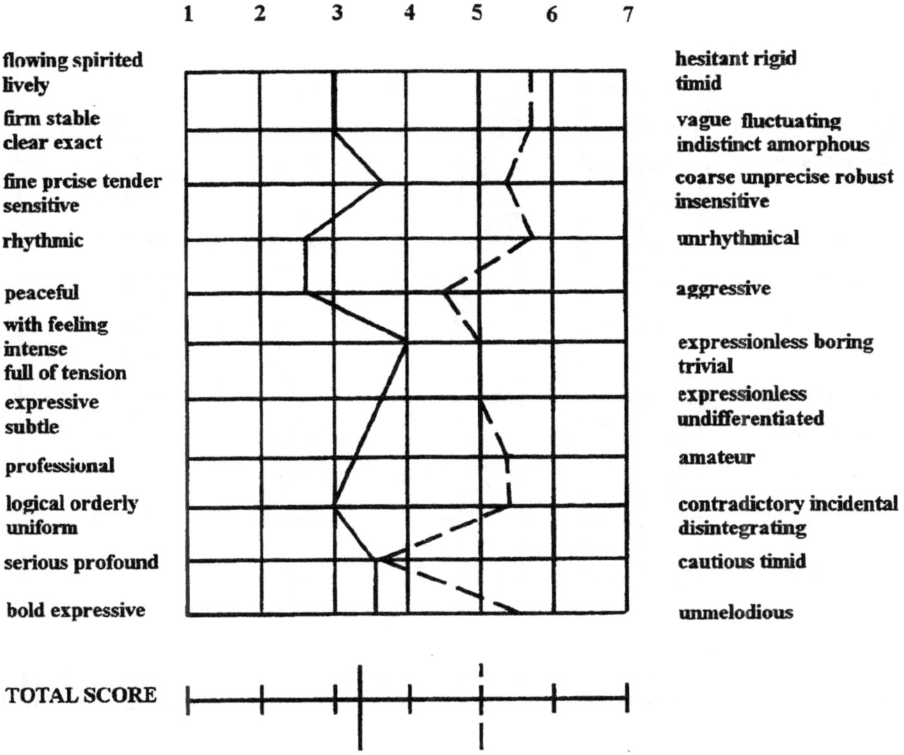

Fig. 5. Polarity profiles of a Händel violin sonata played by the same patient as of fig. 1a,b at U_1 (BPRS 49, *dashed line*) and U_2 (BPRS 22, *solid line*). Means of the ratings of three experts

remitted state were to account for a more apropriate performance. Therefore parts of the recordings selected for the assessment of achievement were carefully analyzed by a musically trained physician, who had no information about patients and controls. Twenty-three peculairities were defined, using the metronome and stop watch and cutting the speed in half. As is seen in Table 1, many deviating performance characteristics and pecularities were seen on average. But there has been no systematic influence of psychopathology upon the frequency of playing mistakes. This becomes obvious when ranking the patients according to outcome. The upper third had really remitted at the second recording time (U_2), whereas the lower third did only partly better Than at the first examination U_1. No meaningful difference between the two samples is seen concerning the variation in performance peculiarities. A tendency for fewer hesitations and wrong notes and a significantly poorer, unclear playing, which means, for example, overuse of the pedal, does not really discriminate the two samples. This means that learned motor patterns are fairly stable in a psychically severely altered condition of a mental disease.

With these results in mind an analysis of the music with respect to expressive content within the different mental states seemed much easier. For the assessment of the music a short polarity profile was developed, which proved fairly reliable and valid (Steinberg and Raith 1985b). Figure 5 shows the English version of the

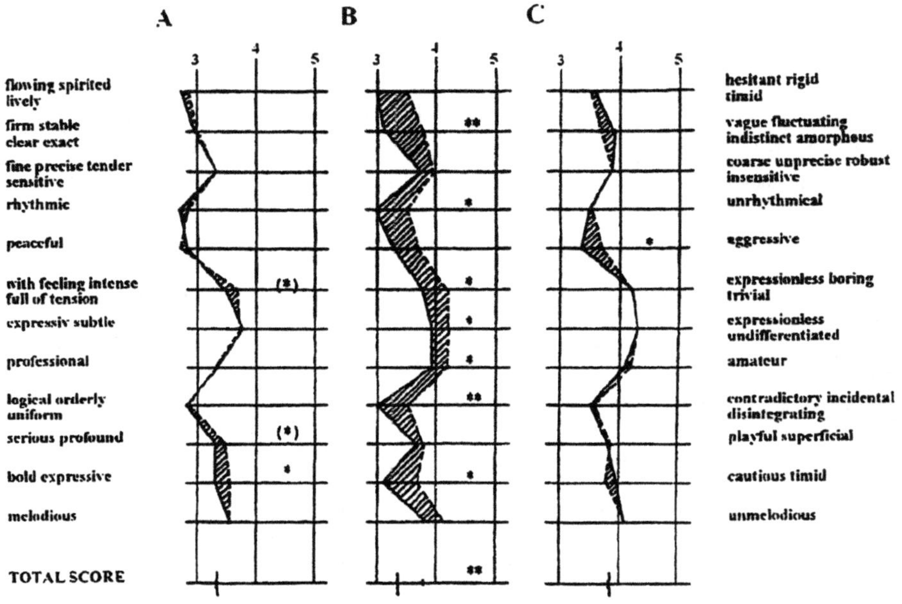

Fig. 6a-c Polarity profiles of literate music pieces played by 29 controls (**a**) [means at U_1 (- - -) and U_2 (——)] (**b**), polarity profiles of the upper third ($n = 20$) of the patients, ranked according to delta-BPRS. **c** Polarity profiles of the lower third ($n = 20$) of the patients. For p values see Fig. 1 (Steinberg et al. 1985)

profile. Single terms of opposite meaning such as melodious - unmelodious or related terms covering an area of musical meaning such as fine, precise, tender, sensitive as opposed to coarse, imprecise, robust, insensitive are divided into a seven step scale. The ratings of three experts were surprisingly stable over time, as were the ratings of 27 controls after 1 year, which also proved for representativity of the experts.

In many cases there proved to be a clear influence of psychopathologic alteration on musical performance (Steinberg et al. 1985), as shown in Fig. 5. The violin teacher suffering from a schizodepressive psychosis (same subject as in Fig. 1a,b) played a violin sonata by Händel considerably worse shortly after admission than 7 weeks later, before discharge. The weakening mainly of the "motoricity" items 1-4 is obvious, as is regularly seen with melancholia. Also a considerable shift in item 9 took place, which is characteristic of schizophrenic symptomatology. As is shown in Fig. 6, patients who remitted completely (Fig. 6b) generally played considerably better than those whose psychopathologic condition did not change very much (Fig. 6c). The latter were almost identically rated at both examinations, as was seen with the controls (Fig. 6a), who were also recorded twice within 6 weeks. But on average controls did more than a half point better than acutely ill patients. These results also show that expressive musical qualities in the average population do not change within a 6-week period of time, unless a severe psychiatric disease intervenes. Also some profile patterns characteristic for specific syndromes or even diagnostic

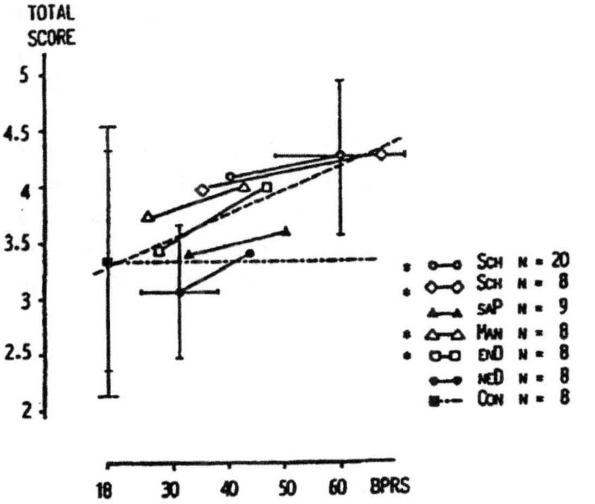

Fig. 7. Means of total scores of the polarity profiles for different nosologic entities at U_1 and U_2. The *dashed line* shows the slope of the regression calculated for all 79 patients at U_1 (BPRS 50.8 11., total score 3.92 .79) and U_2 (33.2 8.7; 3.68 .78). The thickness of the *horizontal line* represents the difference between U_1 and U_2 scores of 29 controls. *sch*, schizophrenia; *sap*, schizoaffective psychosis; *man*, mania; *end*, endogenous depression; *ned*, neurotic depression; *con*, controls. see text

entities resulted. The performance of endogenously depressed patients was much weaker than that of neurotic-depressed patients, despite both groups comparing well as to the extent of depressive syndromes and average musical skills. Manic patients showed the least consistent profile pattern in relation to psychopathology. Endogenously depressed patients again showed a weakening of motoric aspects as with the tempo study (see Figs. 1-4), whereas neurotic-depressed patients were not affected at all in respect to motoricity. Schizophrenic patients in remission did much better in U_2 in polarities, which define logic, appropriate, therefore also rather professional aspects of a performance. This means that performing music reveals different impairments according to different psychiatric diseases, allowing one to speak of music psychopathology.

The construction of the profile allows for a summing up of the items. Figure 7 shows the slope between the average scores at U_1 and U_2 of different nosologic entities. Generally no mental disturbance ameliorates musical achievement. In contrast, growing psychopathologic alteration generally worsens musical expression, independent of the kind of the disturbance, whereas the sum scores of the mentally sane controls result in U_2 almost exactly to the same amount. Whether the steeper slopes of depressed patients in contrast to those of schizophrenic and schizoaffective patients mean a stronger involvement of the musical sense in psychopathology, remains unclear, since no linearity of the items of the Brief Psychiatric Rating Scale (BPRS; Overall and Gorham, 1962) scale can be proven. But the regression line calculated from all U_1 and U_2 data of the patients meets at the zero point of the BPRS scale perfectly, the controls´ levels of achievement confirming the assumption of it being the average achievement score.

By thinking in terms of the results of somatic correlates, a hypothesis of mental disease as a reversible weakening of cerebral functioning was established which is accompanied by a reversible amusic disturbance as, for example, some neurologic diseases. Results of an adapted version of Seashore´s test showed a considerable weakening of some fundamental musical abilities (Steinberg et al. 1989). Schizophrenic patients demonstrated a reversible impairment of reproducing tones, other patient groups and controls, of course, did not. Endogenously depressed patients again proved to have weakened motoricity; reproducing of rhythms was reversibly impaired to a considerable extent. Theses results were not due to a weakened concentration and attention, as was defined with d2 testing (Brickenkamp 1962). Of course the different diagnostic groups were also tested for musical ability.

Whether a gating process in schizophrenia or the global lack of impulsion in endogenous depression causes this, cannot be answered. But all studies mentioned here support the hypothesis of music being an appropriate means of examining cognition, affect and motoricity in mental diseases. Nevertheless the results with literate music remain at the level of musical expression. A deeper analysis is needed to define what conveys the alterations in expressive qualities, which obviously occur "between the notes."

References

Brickenkamp R (1962) Test d2. Aufmerksamkeits - Belastungstest. Handanweisung. Hogrefe, Göttingen

Günther W, Steinberg R, Petsch R, Moser E, Günther R, Streck P, Heller H (1988) Hemisphärenfunktionsstörungen bei Schizophrenen während motorischer und musikalischer Aktivierung, gemessen mit EEG-Mapping, regionaler Hirndurchblutung und Kernspintomographie. In: Oepen G (ed) Psychiatrie des rechten und linken Gehirns. Deutscher Ärzte-Verlag, Cologne, pp 134-146

Günther W, Steinberg R, Streck P, Banquet JP, Bscheid I, Raith L,Stiltz I (1991) Brain dysfunction in psychiatric patients during music perception measured by EEG mapping. Relations to motor dysfunction and influence of neuroleptic drugs. Eur Neuropsychopharmacol 1:143-155

Günther W, Klages U, Mayr M, Haag C, Riedel R, Hantschk I, Streck P, Steinberg R, Baghai T, Banquet JP, Rondot P (1993) EEG mapping investigations of psychomotor and music perception brain dysfunctions in untreated schizophrenic patients. Neurophysiol Clin 23:516-528

Hamilton M (1960) A rating scale for depression. J Neurol Neurosurg Psychiatry 23:56

Overall JE, Gorham DR (1962) The brief psychiatric rating scale. Psychol Rep 10:799-812

Raith L, Steinberg R, Reitschuster S (1989) Amusia and mental diseases. Musik Tanz Kunstther 2/3:121-126

Steinberg R, Raith L (1985a) Music psychopathology. I. Musical tempo and psychiatric disease. Psychopathology 18:254-264

Steinberg R, Raith L (1985b) Music psychopathology. II. Assessment of musical expression. Psychopathology 18:265-273

Steinberg R, Raith L, Roßnagl G, Eben E (1985) Music psychopathology. III. Musical expression and psychiatric disease. Psychopathology 18:274-285

Steinberg R, Raith L, Günther W (1989) Hearing and perception in different mental states. In: Nielzén S, Olsson O (eds) Structure and perception of electroacoustic sound and music. Excerpta Medica, Amsterdam,pp 106-116

Steinberg R, Kimmig V, Raith L, Günther W, Bogner J, Timmermann T (1991) Music psychopathology. IV. The course of musical expression during music therapy with psychiatric inpatients. Psychopathology 24:121-129

Steinberg R, Fani M, Raith l (1992a) Music psychopathology. V. Objective features of instrumental performance and psychopathology. Psychopathology 25:266-272

Steinberg R, Fani M, Raith L (1992b) Music psychopathology VI. The course of objective instrumental performance characteristics with psychiatric inpatients. Psychopathology 25:273-282

Steinberg R, Günther W, Stiltz I, Rondot P (1992c) EEG-mapping during musical stimulation. Psychomusicology 11:154-169

von Zerssen D (1979) Klinisch-psychiatrische Selbstbeurteilungsbögen. In: Baumann K, Berbalk H, Seidenstücker G (eds) Klinische Psychologie. Trends in Forschung und Praxis. Huber, Bern

Music Perception and Preference in the Course of Mental Diseases

Lydia Raith, Reinhard Steinberg, and Heiner Roth

It is the intention to present some results of a study which investigated whether mental diseases can influence perception and preference of musical meaning. The estimation of musical expression, by which is meant the impression of musical messages, is guided not only by the characteristics of music, its composition or performance, but also by individual factors. Some studies have demonstrated that the connotation depends, for example, on personality traits (Cattel and Anderson, 1953b), present condition, mood, concentration (Cattel and McMichael 1960; Nielzen and Cesarec 1982b) etc., and not least on mental illnesses (Nielzen and Cesarec 1982c; Cattel and McMichael 1960). A psychiatric disease impairs affective, cognitive, motoric or psychomotoric functions, which are especially characterized by the kind of illness. One is compelled to suppose that psychopathological symptoms influence musical processing field.

In 1953 Cattell and Anderson (1953b) started a publication with the words: "The powerful and immediate connection of musical stimulation with emotional experience, ..., have long pointed to measures of musical preference as effective avenues to deeper aspects of personality...". They found correlations between the 16 PF Questionaire and the Musical Preference Test of Personality. Later, in 1960, Cattell and McMichael published results of a study of 230 psychiatric patients. They described correlations between the Musical Preference Test and some diagnostic groups. Patients suffering from paranoia showed high scores in the factor "paranoid imperviousness," schizophrenic patients in the factor "introspectivness", patients with affective psychoses in the dimension "tenacity - cyclothyme relaxedness."

Other studies have investigated music reception or music sensation in connection with mental illnesses. For example, in a study of Nielzen and Cesarec (1982d). 107 psychiatric patients and 100 healthy controls were exposed to seven pieces of music. They rated the music on semantic differential scales describing three pairs of factors: "tension - relaxation, gaiety - gloom, attraction - repulsion". Compared with controls the psychiatric patients, divided into seven diagnostic groups, showed several clearly different experiences of music. The psychopathology seemed to influence the emotional experience of music much more than age and personality factors. For example, schizophrenic patients rated the music more attractive, depressive and manic psychotics more gloomy, depressive and anxiety neurotics less attractive than controls.

Simon et al. (1951) investigated the recognition of happy and sad moods of music. Controls and depressive psychotics made fewer errors in recognizing happy music. Depressives as well as controls disliked sad records, whereas manic and schizophrenic patients liked them.

Method

One aim of the present study was to investigate whether inpatients with mental disease experience the sense of music differently than healthy persons. Another aim was to find out whether the ratings of patients changed during their clinical treatment, when recovering. The participants were exposed to nine pieces of music lasting between 60 and 90 s. We tried to offer several varied pieces of music, classical pieces as well as light music or popular songs from the hitparade.

We offered the following pieces:

- A.Vivaldi, (Four Seasons),"Winter," 1st movement, Allegro non molto
- B.Bartok, Divertimento, 2nd movement, Allegro
- Scott Joplin, "Maple Leaf Rag"
- J.S.Bach, Brandenburg Concerto no.4, 2nd movement, Andante
- Tony Sheridan, "When the Saints Go Marching In"
- L.v.Beethoven, 3rd Symphony, 1st movement
- F.Gulda, Concerto for Violoncello and Wind Ensemble, 5th movement
- Whitney Houston, "All at once"
- W.A.Mozart, Symphony no.29 in A-Major, 1st movement, Allegro moderato

These pieces were to be rated in regard to 21 polarity items of a semantic differential. Some of the items correspond to items of other authors, especially to items by Nielzen and Cesarec (1982a) and by Reinecke (1967,1973).

Inpatients n=96 of the Psychiatric University Hospital Munich and healthy controls n=39 participated in the study. The participants were tested in groups of four to eight persons by two investigators. After giving the patients instructions the music pieces were played over tape; the intervals between pieces took until the last subject had ended his rating.

In addition to the psychiatric diagnoses according to ICD-9 (International Classifikation of Diseases) the psychopathological state was assessed using the Brief Psychiatric Rating Scale (BPRS) authored by Overall and Gorham (1962). The results are divided into five global syndromes: anxiety/depression (ANDP), anergia (ANER), thought disorders (THOT), activation (ACT), and hostility (HOST). The total score with the zeropoint of 18 is a general measurement of psychiatric disorders. On the day of testing the physician in attendance rated the 18 items of the scale.

Groups

The patients were divided into three groups: 41 patients with schizophrenia (SC), 31 patients with neurotic depression (ND), 24 patients with endogenous depression (ED). Fifty-one patients were tested twice, at the beginning of treatment in a state of serious illness and at the end of treatment in a state of at least partial improvement. All controls were retested after about 6 weeks, which corresponds to the time between the tests of the inpatients.

Table 1. Numbers and BPRS scores in each patient group at the time of the test and retest

	SC (n=22)		ND (n=12)		ED (n=17)	
	U1	U2	U1	U2	U1	U2
ANDP M:	9.86	8.32	14.17	11.92	16.12	12.06
S:	4.26	2.95	3.66	5.09	3.12	3.94
SN:	(+)		(+)		++	
ANER M:	11.59	11.36	7.83	6.67	7.71	6.76
S:	3.49	3.13	3.41	2.46	2.49	3.11
SN:			(+)			
THOT M:	9.50	8.23	6.42	5.42	5.00	4.47
S:	3.54	3.29	3.70	1.62	1.62	.94
SN:						
ACT M:	6.14	5.18	7.00	6.25	6.71	4.76
S:	2.34	2.13	3.19	2.30	2.54	2.05
SN:	(+)				+	
HOST M:	7.05	5.59	5.92	5.08	4.71	4.00
S:	2.85	2.56	2.68	2.43	2.26	1.67
SN:	+					
BPRS M:	44.14	38.68	41.42	34.17	40.24	32.06
S:	8.75	2.56	9.75	8.13	7.56	7.94
SN:	+		++			
δ-BPRS	21%		31%		37%	

SC, schizophrenia; ND, neurotic depression; ED, endogenous depression; ANDP; anxiety/depression; ANER, anergia; THoT, thought disorders; ACT, activation; HOST, hostility; BPRS, Brief Psychiatric Rating Scale; M, mean; SD, standard deviation; SN, level of significance
++, $p < 0{,}01$
+, $p < 0{,}05$
(+), $p < 0{,}10$

Table 1 shows the numbers and the BPRS scores of each patient group at the time of test (U1) and retest (U2). Delta BPRS represents the percentage of decreasing of the BPRS total score. The differences were significant in all groups. Remembering the zeropoint of the scale - a score of 18 - the patients did improve but they were not yet totally recovered. The psychopathological symptoms had clearly, although not completely, improved at the time of retesting.

Results

The ratings of eight pieces of music could be interpreted. The first one, Vivaldi's "Winter," has to be excluded, because some of the participants did not seem to be familiar with the kind of rating yet. For example, either they had chosen only the extremes or they hadn't filled out the whole list.

First the results of the interindividual design are presented. The comparision of the three patient groups and the control group should answer the question of whether there are statistically significant differences between the groups.

On the whole the ratings of the four groups show correlation to a great extent. The averages of the profile items take a similar trend, nevertheless there are some evident differences between the patient- and the control-group. The eight pieces of music were rated differently, and the significant differences between the groups

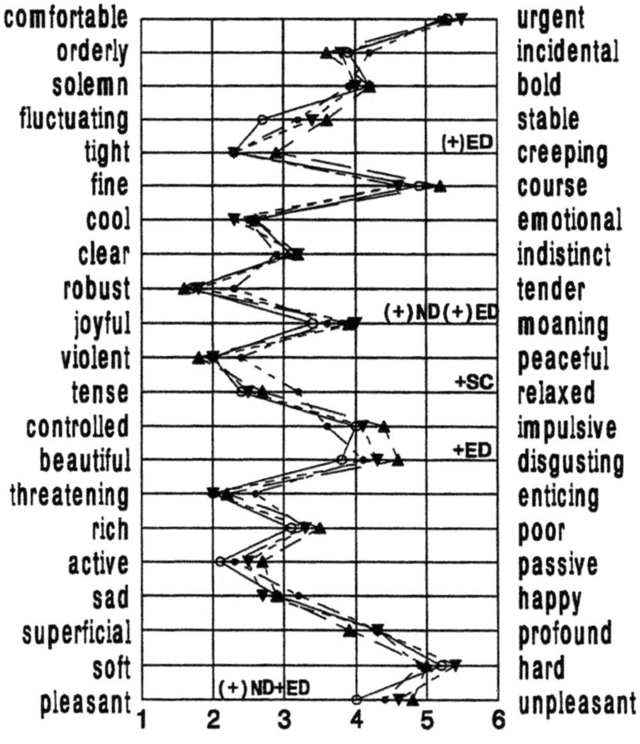

Fig. 1. B.Bartok, Divertimento, 2nd movement, Allegro, Control group (*KG; n* = 39); schizophrenia (SC; *n* = 41) neurotic depression (*ND; n* = 31); endogenous depression (*ED; n* = 24). ○ *KG*; ● *SC*; ▼ *ND*; ▲ *ED*

also varied. We have selected the profiles of two pieces, because no systematic trends in group differences could be found. On the whole, the pieces were judged quite similarly by the four groups.

Figure 1 shows the profile of the music piece which received the most similar rating, - Bartok "Divertimento." The mean scores and the level of significance of differences between a patient group and the controls is marked on the right side of the scale. Except for the two trends concerning the items tight - creeping and joyful - moaning, evident differences could only be found in 3 items. Compared with the control group patients with schizophrenia rated the music as less tense, patients with endogenous depression as less beautiful. Neurotic and psychotic depressives disliked it more than healthy persons.

Figure 2 demonstrates the profile of Mozart's Symphony in A-major. The estimations seem mainly to be influenced by a depressive syndrome or a depressive affect. Compared with healthy persons patients suffering from endogenous as well as from neurotic depression diverge. Above all the ratings of neurotic depressives differ from those of controls, although some of the levels of significance reach only 10% and could be interpreted only as a trend. Neurotic depressives perceived the music as less emotional, as more robust, as less peaceful, relaxed, beautiful, enticing and poor. Psychotic depressives rated it also as less emotional and as less joyful. Compared with controls all diagnostic groups judged the music as less beautiful and as less pleasant. Among the three patient groups two significant differ-

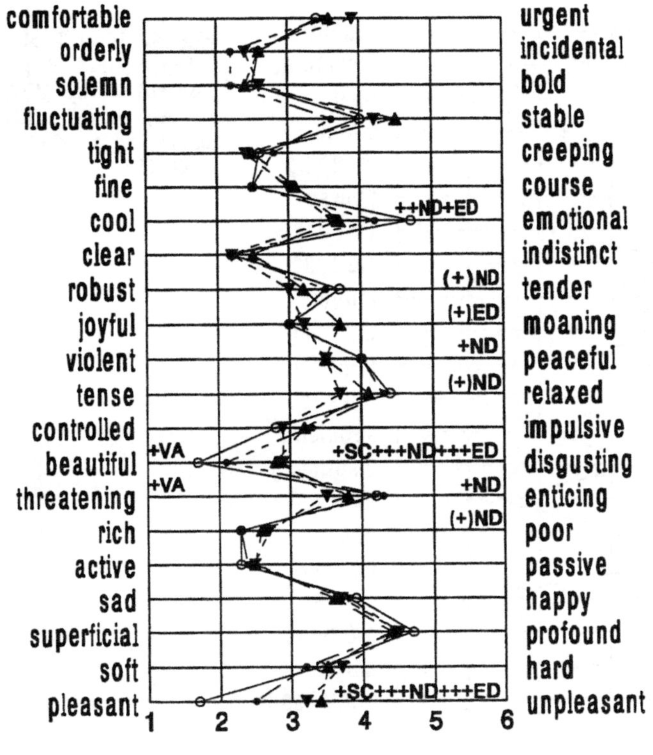

Fig. 2. W.A. Mozart, Symphony No. 29, A-major, 1st movement, Allegro moderato, Control group (*KG*; $n = 39$); schizophrenia (*SC*; $n = 41$) neurotic depression (*ND*; $n = 31$); endogenous depression (*ED*; $n = 24$). *VA*, Varianceanalysis; ○ *KG*; ● *SC*; ▼ *ND*; ▲ *ED*

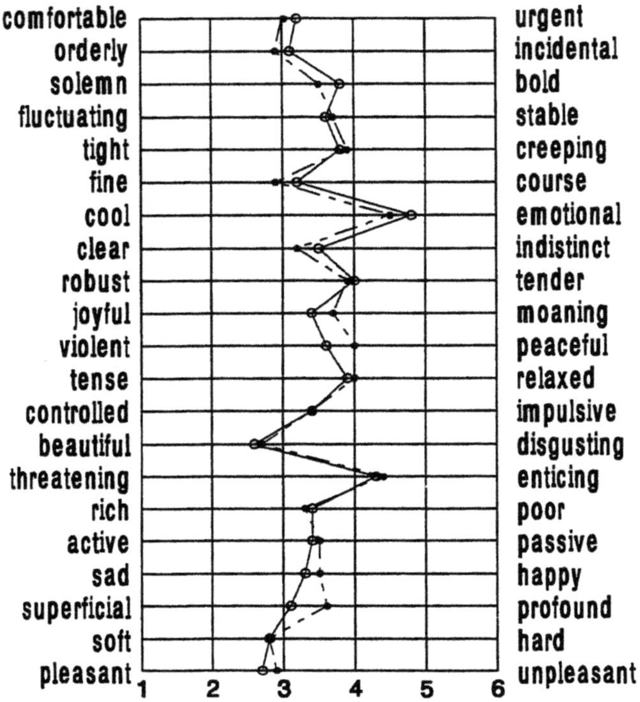

Fig. 3. Whitney Houston "All at once." Control-group (*KG; n* = 39); neans at U1 and U2; ○ *KG* U1; ● *KG* U2

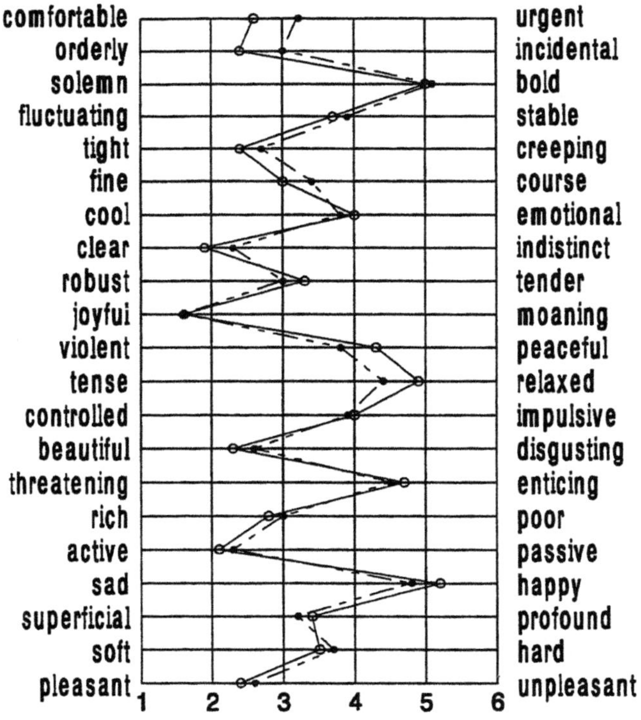

Fig. 4. Scott Joplin "Maple Leaf Rag." Control-group (*KG; n* = 39); means at U1 and U2; ○ *KG* U1; ● *KG* U2

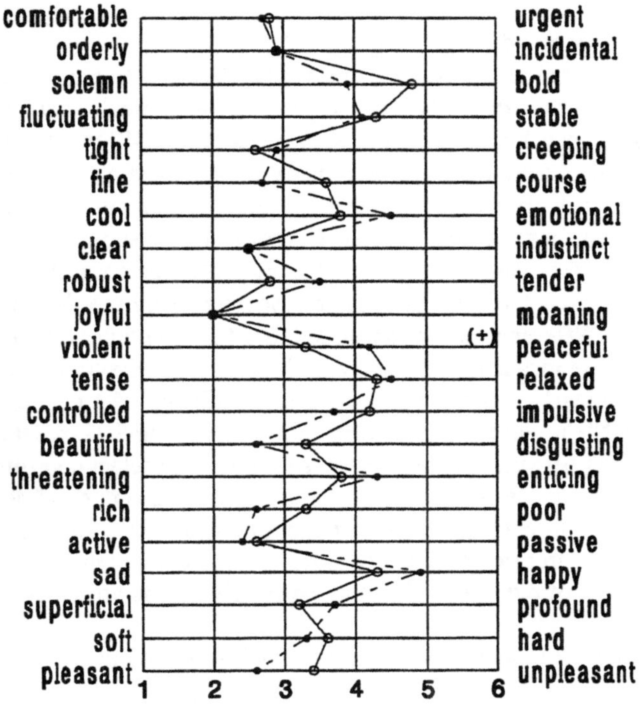

Fig. 5. Scott Joplin "Maple Leaf Rag." Neurotic depression (*ND*; $n = 12$); means at U1 and U2; ○ *ND* U1; ● *ND* U2

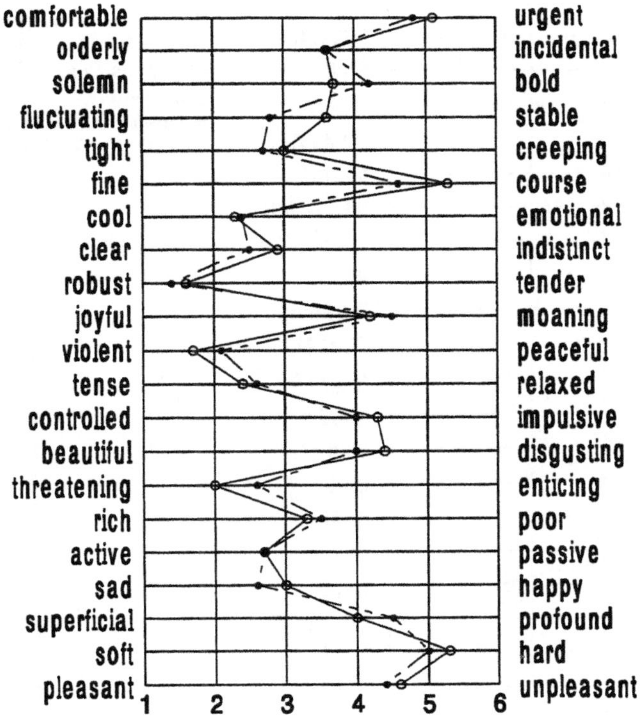

Fig. 6. B. Bartok, Divertimento, 2nd movement, Allegro. Endogenous depression (*ED*; $n = 17$); means at U1 and U2; ○ *ED* U1; ● *ED* U2

ences were found, which are noted on the left side of the scale, concerning the polarities beautiful - disgusting and threatening - enticing.

As to the second part of the study, the intraindividual design was chosen to answer the question of whether the impression or the connotation of music changes during clinical treatment of psychiatric patients, when recovering. Differences in ratings between test and retest were investigated in the three diagnostic groups as well as in the control group. On the whole there were fewer significant differences between test and retest than between the groups. Contrary to our suppositions, there were some changes in the ratings of controls. The piece of music which was estimated very consistently was the song by Whitney Houston. Figure 3 demonstrates the only significant difference between test and retest in the polarity cool - emotional. In this profile the means range mainly in the middle of the scale; the participants didn't decide clearly between the contrasting attributes. The next example, Fig. 4, was the profile to Scott Joplin's "Maple Leaf Rag," which shows the most significant differences between test and retest. At the time of retest controls rated the music tendentially as less comfortable, evidently as less orderly, fine, peaceful, relaxed and happy. In contrast to the last example this music was more clearly judged; the average scores range from 2 to 5. It is interesting that the scores in retest move to the middle of the scale. This may be interpreted as an effect of adaptation.

Now to the results in the patient groups. Contrary to our expectations no sys-

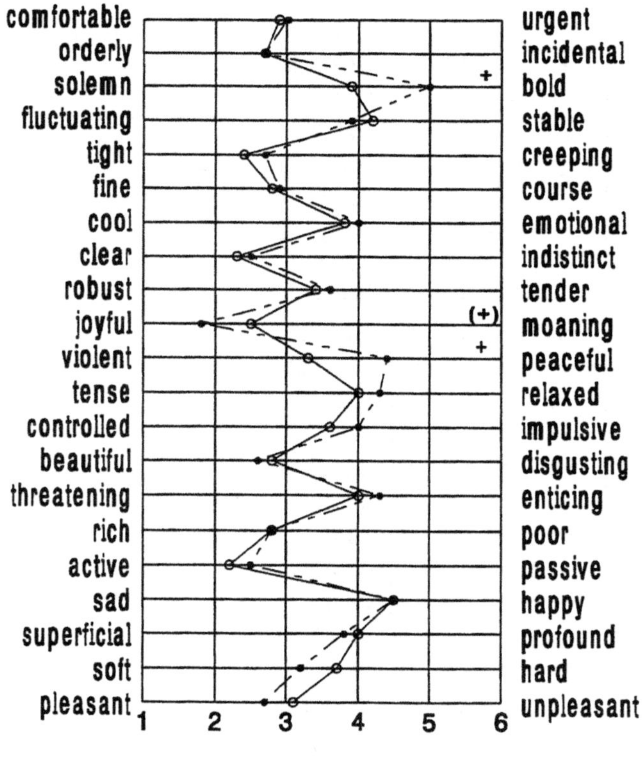

Fig. 7. Scott Joplin "Maple Leaf Rag." Endogenous depression (*ED; n* = 17); means at U1 and U2; ○ *ED* U1; ● *ED* U2

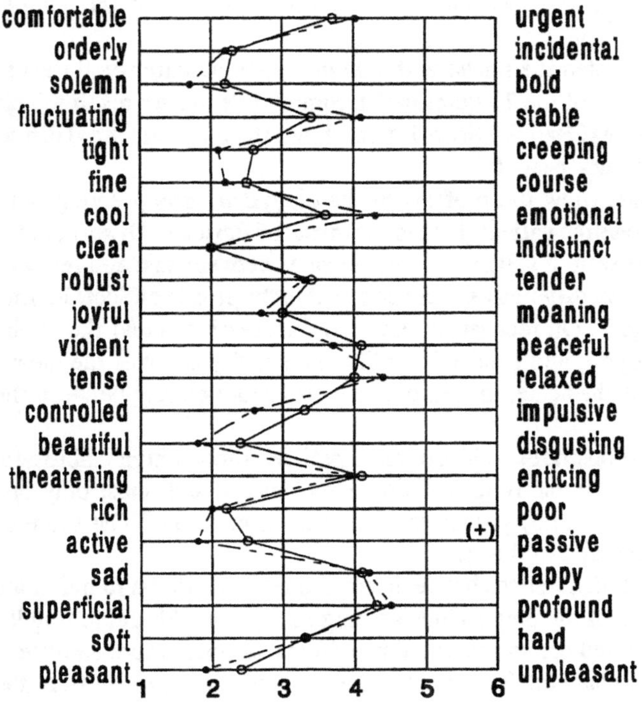

Fig. 8. W.A. Mozart, Symphony No. 29, A-major, 1st movement, Allegro moderato. Schizophrenia (SC; $n = 22$); means at U1 and U2; O *SC* U1; ● *SC* U2

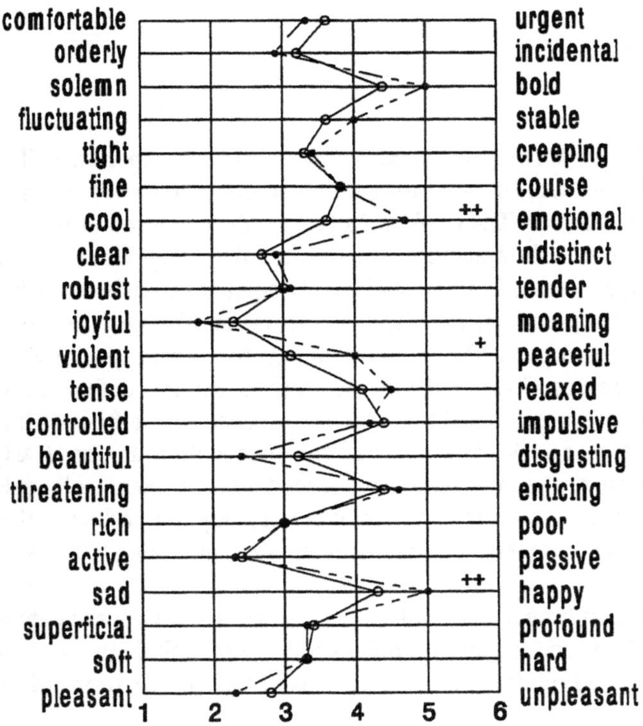

Fig. 9. Tony Sheridan and the Beat Brothers "When the Saints Go Marching In." Schizophrenia (SC; $n = 22$); means at U1 and U2; O *SC* U1; ● *SC* U2

tematic differences were found between test and retest. On the whole the number of differences was low.

Neurotic depressives rated the music with a high degree of stability, for every piece of music at most two of the 21 items differ. Figure 5 demonstrates this. This example, again "Maple Leaf Rag" by Scott Joplin, shows the only tendential difference in the polarity violent - peaceful.

The following figures show the profiles of the diagnostic group endogenous depression. As Fig. 6 shows the patients judged the music of Bartok's "Divertimento" quite consistently; there were no significant differences between test and retest.

The most intraindividual differences could be found in the profile of Scott Joplin's "Maple Leaf Rag" (Fig.7). On recovering psychotic depressives rated the music clearly as more bold, joyful, and peaceful. Unlike the healthy controls the means move to the extremes of the scale; the patients decided more clearly between the contrasting attributes.

Finally, as Fig. 8 demonstrates, schizophrenic patients show a high stability in their rating of Mozart's Symphony in A-major. Between test and retest only one tendential difference exists. They judged the music as more active. The score of retest approaches the extreme of the scale.

The most and clearest differences between test and retest could be found in the following profile of "When the Saints Go Marching in" by Tony Sheridan (Fig.9). Schizophrenic patients rated the song at the time of retest as much more emotional and happy, and as more peaceful. The mean scores of retest again move from the middle of the scale.

Conclusions

Combining these results we found great consistency in musical experience. The diverse musical pieces were judged quite differently according to the 21 items of the semantic differential we used. The profiles describing the mean scores of each patient group and the controls show an approximately similar trend. However, the comparison of the groups resulted in some statistically significant differences between each diagnostic group and the healthy controls. Generally the psychiatric patients rated the music more in the direction of the negative side, especially patients with depressive illnesses. Compared with the controls, patients with neurotic as well as with psychotic depression disliked five and patients with schizophrenia three of the eight musical pieces. According to this they judged the music as clearly less beautiful. In this context it seems worth mentioning that an influence of music listening habits was controlled too. The participants in the study were asked wether they like listening to music. Significant differences between the four groups were not found.

The results of the intraindiviual design also demonstrate a great consistency in the music perception. The profiles were still quite similar, although there are some differences between test and retest. Contrary to our expectations, the most differences were found in the control group. All significant differences tended move to the middle of the scale at the time of retesting. This may represent an effect of adaptation - when listening to music the second time, the element of novelty has worn off.

Patients suffering from neurotic illness rated at most two items significantly different when recovering. More evident differences between test and retest could be found in the groups of psychotic depressives and schizophrenic patients. However, at most the mean scores of three items changed. In contrast to the healthy controls patients suffering from endogenous depression and from schizophrenia rated more extremely at the time of retesting. They decided more clearly between the polarities during recovery An effect of adaptation could not be found.

On the whole, it seems that psychiatric disease and its improvement influences music perception and preference less than supposed. On the other hand it must be considered that patients in a very severe state of illness could not be tested, and that patients with manic disorders could not be included either. The participants had to be able to concentrate during the session, which took about 45 min. We must also take into account that the patients were not completely recovered at the time of retest. The BPRS scores show that they suffered from resting symptoms, although their psychopathological symptoms were improved.

Finally the small number of the patients must be taken into consideration, and we could not analyze the effect of special psychopathological symptoms within diagnostic subgroups. The effect of medication could not be systematically analyzed either; because of the individual selection and dosage of drugs with their various effects there is great heterogeneity between and within the groups. In an earlier study investigating the musical tempo an influence of medication could not be found (Steinberg and Raith 1985).

Concerning the controls, who could not be assessed with the BPRS, a short attribute list was used to assess present condition and mood. The statistical evaluation is not yet completed.

The results of the present study seem to allow the conclusion that a great inter- and intraindividual consistency in musical experience exists. Even a severe mental illness could not influence the perception and the rating of music on principle. These results do not seem to be compatible with the supposition that there is a relationship between musical preference and individual factors or personality traits, which, for example, Cattell and McMichael (1960) have demonstrated. Musical experience seems to depend basically on the characteristics of music.

On the other hand, the few differences between test and retest in healthy persons indicate an effect of adaptation which could not be found in psychiatric patients. Especially psychotic patients seem to follow a reverse trend. When recovering they gained more clear perception of some musical aspects; they decided more clearly between contrasting attributes. It may be supposed that severe mental disorders reduce musical experience. Up to now hardly any systematic follow-up studies exist. So we can't compare our results directly with the results of other studies.

References

Cattell RB, Anderson JC (1953a) The IPAT music preference test of personality. IPAT, Champaign Ill

Cattell RB, Anderson JC (1953b) The measurement of personality and behaviour disorders by the IPAT music preference test. J Appl Psychol 37: 476-454

Cattell RB, McMichael RE (1960) Clinical diagnosis by the IPAT music preference test. J Consult Psychol 24(4):333-341

De la Motte-Haber(1985) Handbuch der Musikpsychologie. Laaber, Laaber

Kleinen G(1968) Experiemtelle Studien zum musikalischen Ausdruck. PhD thesis University of Hamburg

Nielzen S, Cesarec Z (1982a) On the perception of emotional meaning in music Psychol Music 9 :17-31

Nielzen S, Cesarec Z (1982b) Emotional experience of music as a function of musical structure. Psychol Music 10: 7-17

Nielzen S, Cesarec Z (1982c) Aspects of tempo and perception of music in mania. Acta Psychiat Scand 65: 81-85

Nielzen S, Cesarec Z (1982d) Emotional experience of music by psychiatric patients compared with normal subjects. Acta Psychiat Scand65: 450-460

Osgood CE (1952) The nature and measurement of meaning. Psychol Bull 49: 197

Overall JE, Gorham DR(1962) The brief psychiatric rating scale. Psychol Rep 10: 799-812

Reinecke HP (1967) Über Allgemein-Vorstellungen von der Musik. Eine experimentelle Untersuchung musikalischer Stereotype mit der Methode des Polaritätsprofils. In: Finscher L, Mahling CH (ed) Festschrift für W.Wiora. Bärenreiter, Kassel, pp 31-40

Reinecke HP (1973) Einige Überlegungen zum Begriff des musikalischen Verstehens. Volk, Cologue

Simon B, Holzberg JD, Alessi SL, Garrity DA (1951) The recognition and acceptance of mood in music by psychotic patients. J Ment Nerv Dis.114:66-78

Steinberg R, Raith L (1985) Music psychopathology. I. Musical tempo and psychiatric disease. Psychopathology 18:254-264

The Reality of Musical Delusions

Reinhard Steinberg

In comparison with the visual arts, music is rarely the kind of mental patients tend to express themselves in. Maybe the argument also holds here that in the general population fewer individuals have a musical aptitude, above all compositional skills, than have visual expressive gifts. Therefore musical material is only exceptionally incorporated in delusional ideas or delusional systems, and a common description or classification system can hardly have been established. But there are some other reasons inherent to music as an emotional language which prevent systematic classification of the musical delusional material comparable to thought contents. Some ideas emerging from this exciting field will be presented.

In the literature there are only a few descriptions of musical delusion which, together with one's own case material, are restricted almost exclusively to schizophrenia. Other mental diseases such as mania, depression, and all organic cerebral disturbances seem scarcely prone to musical delusions. Generally, delusional thinking is characterized by a false belief that is firmly held, despite objective and obvious contradictory proof or evidence and despite the fact that other members of the culture do not share the belief (Kaplan and Sadock 1991). Psychiatry classifies delusions according to prevailing ideas as religious, somatic, paranoid, or grandiose delusions. But more important as regards therapeutic outcome seems to be whether delusional thinking is bizarre, incomplete, unorganized, or even systematized, the latter seeming to have a poorer outcome.

In schizophrenia musical delusional processing can be roughly divided into three categories (Table 1). A certain group of patients, characterized by unsystematic concrete thinking and an inadequate and parathymic affect, likes their musical productions - which some of the patients call compositions, others improvisations - to be heard. Those patients are not usually trained or especially musically gifted. For example, a 52-year-old housewife with grandiose ideas begged to be heard and for her piano improvisation to be recorded. Never musically educated, she had started years before, shortly after the beginning of the first schizophrenic episode, practicing piano for hours a day, which was barely tolerable for her family. Figure 1a shows in a metrically not completely correct notation the "harmony tones," "question harmony tones," and "joy studies," terms which reflect her distorted thinking and the schizophasic neologisms, which were also characterized by a manieristic pronounciation. Figure 1b shows another example of an improvisation she exercised daily "for hours" at home, which takes a rhapsodic form representing a musical gesture. Another characteristic of this piece was a preference for the augmented fourth, as seen in Fig. 1c, suggesting French impressionistic music. She called them "question harmonies," which were played softly with overuse of the

Table 1. Three types of musical delusions, the prevailing psychopathologic state, and its correlation with musical form

Type	Musical form	Pathological state
I	No musical form, rather tinkling; perseverations of simple motifs	Elated affect; manic or schizophrenic, preferentially hebephrenic patients without significant musical training
II	Delusions derived from simply structured musical material, which represents "world harmonies, cosmic power" e.g.	Paranoic/schizophrenic patients with systemathised delusions of grandeur. Often some musical formation
III	Real skilled compositions with a paranoic connotation, mostly systematized	Manic or schizophrenic patients with a good musical formation

pedal harmonies and frictions. When listening to the recordings, which in some ways were very interesting, the unusual meaning or connotation, of course, can´t be heard.

Altogether, she did much better than five other hebephrenic patients, whose "composing" on the piano was boring, persevering in playing scales or short motifs for minutes or even longer. The distressing behavior of a young hebephrenic is illustrated in Fig. 1d-e, where a 10-tone scale as well as a "parallelogram" was repeated stereotypically for a quarter of an hour, until the recording psychiatrist lost his patience. The shape of a parallelogram really arises, when drawing lines between the white and black keys of the motif. The endless playing of scales was accompanied by the patient counting in a low voice "one, two, three,..., nine, ten, ten, nine,...two, one, one, two,...". Although he had great music in his mind, he never revealed his delusional thinking to his physicians.

In contrast to this stereotypic playing, some patients lack rhythmicity, persist in changing notes, and are prone to a strange practice of placing both hands on the middle of a key in a slow, screwing manner moving towards opposite ends. Obviously a technical aspect of the instrument has become a matter of interest too, however , the music always means something important to the patients, sometimes systematically being incorporated in delusional ideas as well.

A second category (Table 1) of musical delusions comprises systematizations which often derive from rather simply structured musical material such as an ending cadence, by relating them to nonmusical theories or ideologies, expanding in part

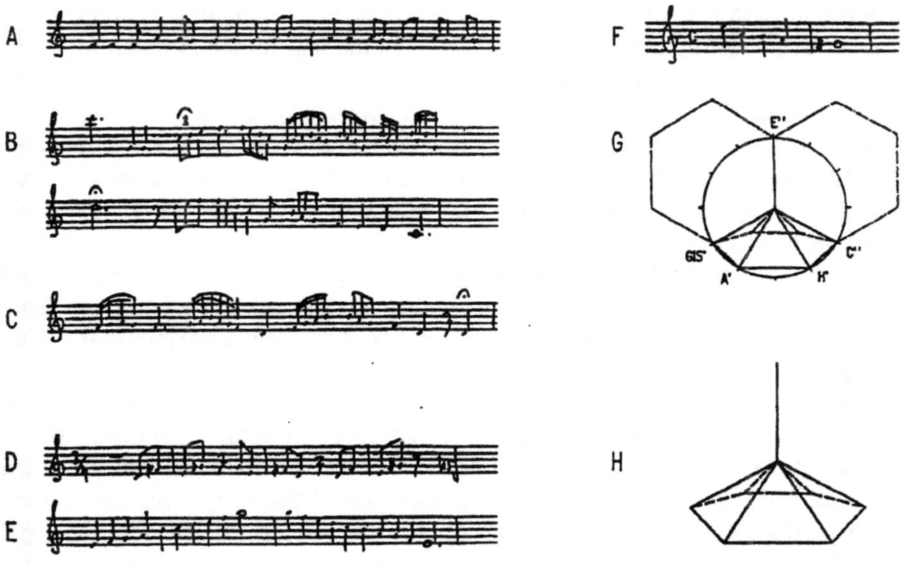

Fig. 1. a-c Music "composed" by a schizophasic patient. **d-e** A short motif ("parallelogram") and a rising and falling scale, repeated stereotypically for minutes by a young hebephrenic. **f-h** A systematized musical delusion, imaginated by a young schizophrenic. see text

to cosmic dimensions. Figure 1f-h shows one of those delusions; it was developed by a 40-year-old academic who suffered for 10 years from episodic paranoic schizophrenia. As a gifted violinist he identified himself in a paranoic manner with one of the greatest Italian masters. In his apartment he had plasterd the walls with a dozen of layers of journals, representing the circles of his world theater, with himself holding court in the middle of important musicians, politicians, scientists, philosophers and so on. But the sytematized musical delusion he developed seemed at first glance only loosely connected with this megalomania. Years previously in a "*coup de foudre*" he had recognized that the whole world with all its rules and physical laws is derived from the keys of an incomplete A-minor cadence. He putted the notes in the graphic transformation of the 12 tones of the pitch-class circle and connected them with the focal point. He wrongly imagines the appearance of a hexagon, which can only be understood as equidistant, when leaving the two-dimensional projection of the 12-tone circle, and at the same time, violating the three-dimensional rules of technical drawing. From this primordial delusional cell - the author spoke of a crystal or atomic structure - an infinite number of hexagons can be derived, representing an infinite number of mystic harmonies never heard on earth. Although the author of this grandiose delusion did not really define his position within this harmonic system, he indicated the identity of the focal point within the 12-tone circle with his position in his world theater. At the piano he gave also an example of the infinite harmonies deriving from this point, which naturally resulted in a poor imitation of his magnific delusion. This kind of delusional systematization of musical material is not so seldom (Steinberg 1987; in der Beeck 1979, 1987). Whereas both melodic and rhythmic material seems more to be incorporated in concepts of self-reference, the harmonic relations are often expanded to complete systems of world philosophy and eternity, which in its concretized form prove them phantastic and paranoic, but often remain externalized. That the author of such a systematized delusion is incorporated as the central point, that he himself is the center, is the clandestine or ouvert aim of all delusions, whether musical or not.

A third group of musical delusions (Table 1) comprises real musical compositions worked out in detail according to the individual competence of the composer. These pieces represent a strange meaning or even a veritable delusional program. Figure 2 shows parts of a composition for violin. The whole sonata consisting of six movements with a duration of 17 min depicts a space flight in the fifteenth century. Unusual namings of the parts, e.g., "Cum angeli et archangeli - spiritui sancti - quasi con balaleica - bebop - litanei deciso e dolce" for the third movement are even exceeded by the explanations the author gave of his compositional program. "...Now they eat, they eat their plant pills in the spaceship, you hear them eat noisily...Emilia, she is the wife of Carol Wojtila, the pope...These notes here put the question, whether he is pure and chaste?... These notes here mean the vicar of

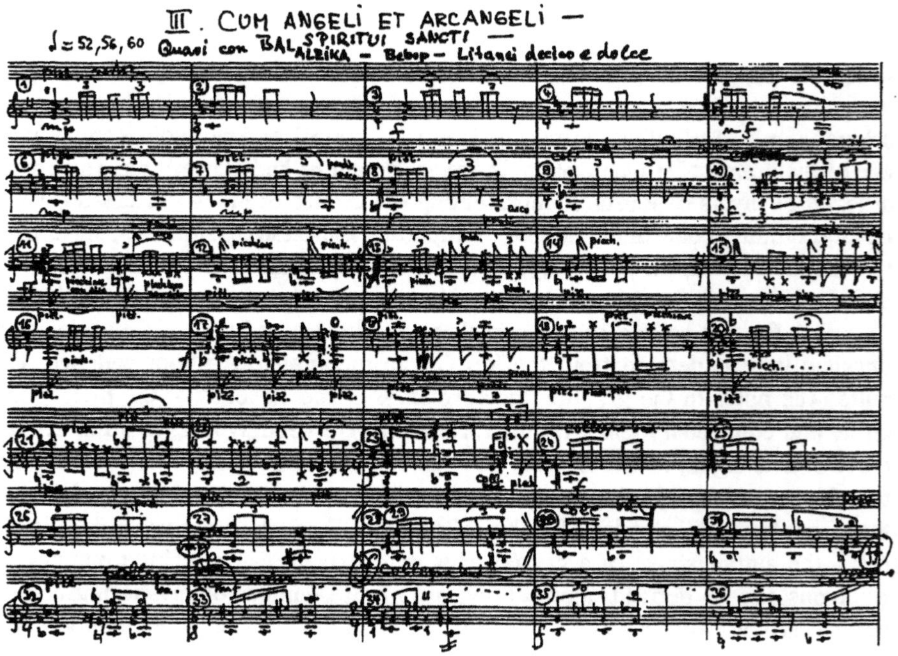

Fig. 2. First bars of the 3rd movement of a violine sonata with a systematized delusional content. see text

Fig. 3. Parts of a piece for piano composed by a music student. Although the writing seems still competent, the extensive behavior due to his schizomanic psychosis is already seen by such picularities as "fffff" followed by "ppp" (Pethö 1967)

Fig. 4. Strange scribblings of musical and nonmusical symbols, representing a composition by a schizomanic music student. At the *bottom left* and *right* a condensation of music graphoelements relates his megalomanic delusion (Pethö 1967)

Christ, he stands on his head... The spaceship moves off, until earth becomes a small, high G flat, four octaves above middle G..." Some notes gain a concrete meaning, as for instance the scanning of "Jesus from Nazareth" or "Ludwig van Beethoven". At the end of third movement G means earth, G flat means underworld, and G sharp is Venus. Many other explanations show that the whole composition represents a complete paranoic system, in which religious as well as sexual ideas are confounded in a phantastic private world. At the time of his explanation the composer, who had retired early and received a pension from his former position as violonist in a renowned orchestra, did well but was still convinced of the reality of this story, a behavior German psychiatry describes with the term of double entry book-keeping.

But when listening to his own recording, one hears violin solo music being played competently by an excellently trained professional musician, but not a delusional system at all. The systematized musical "Wahn" needs verbal explanation as does all absolute music which we connect to a program of semantic meaning. Formally, his composition contains elements of twelve-tone and serial music. A noticeable feature is a grouping into two bar phrases using fermatas over rests, tremoli fading away in harmonics or pizzicati of most extreme pitches. The structure is overcomplicated musically, consisting of abrupt jumps in extreme pitches and rapidly changing extreme dynamics within very short periods of time. Also, for expressive purposes the violin is being used in ways in which it was not built for originally, such as *col legno* playing, knocking on the violin's body, wiping over the chords. His competence as a trained composer is also seen in the script. But concerning the form, the movements become successively shorter, ending with the sixth movement in a two bar composition. This sonata has not been published up to now; however, with some other compositions the composer has been successful to some extent.

In der Beeck (1987) published parts of an opera composed by a young paranoic student of composition, which did not reach the artistic standard of the violin sonata. Pethö (1980) followed up a mental patient, a former student of composition, for 16 years, documenting an obvious decline in compositional competence until all the musical intention of the composer was expressed in a single complex chord. The former melodic imaginativeness, the structural and dynamic competence as well as his artistic dicursiveness had become desolate, coincidentally with his change in personality due to the chronic course of his schizophrenic psychosis. The change

Fig. 5. Drawing with "compositions" by Adolf Wölfli, a schizoprenic inpatient in a Swiss asylum. It is entitled "Schimpanzee - Monkeys of the Union Canada: America" (Spoerri and Glaesemer 1967)

from competent writing (Fig. 3) into scribbling (Fig. 4) coincides with the loss in musical competence. This music, however, is not marked by the mental disease as much as with Adolf Wölfli´s drawings, which very often contain musical material (Fig. 5). The music is a copy of parts of the songbook used in the Swiss asylum Wölfli lived for more than 40 years (Streiff and Keller 1976).

Discussion

When listening to the violin sonata, most people will certainly have found the music strange, but in some aspects attractive and meaningful. Without the explanation of the composer, it would never be known that the motivating force for composing the violin sonata was a complete delusional sytem. Music can not relate a delusion or transmit any semantic meaning, neither a sane nor a delusional one. Of course we understand the meaning of averbal music as, for instance, national anthems, funeral marches, Bavarian yodellers, and Viennese waltz. Our understanding of music is associated with concrete events, things, meanings, or terms to certain melodies, harmonies and rhythms, but music never means the concrete concept in semantic terms. And so music has no reality or semantic logic comparable to dreaming states. Only in a delusion can it achieve an otherwise forbidden reality of delusional certainty.

One could, of course, argue that for many people certain music has a special meaning which may even pertain to the proof of the existence of God. We can be sure that Bach, Bruckner, and Olivier Messiaen would insist on their beliefs. But they would readily admit, too, that their music represents their convictions, but is not meant to depict a concrete person or is more a symbol of God, taking, for instance, the three-quarter measure as a symbol of the Holy Trinity. The loss of ability to understand symbols is one of the central mechanisms underlying delusional thinking. Regarding program music, the program is never semantically concrete, but rather it symbolizes its content. Former attempts at concreteness, such as Kircher´s doctrine of affections, did not survive the attempt to concretize musical affects. More recent attempts to explain Mozart´s music by concretistic derivations from the musical texture seem at the least incorrect and confused (Born 1982).

But what is strange about the music of these amateur and professional composers. Formally, they violate the rules by which music is played, improvised, or composed within the norms of society. Persistence as shown with the "parallelogram" is boring and annoying; it is also an unscrupulousness to produce unstructured, ruleless musics without end. Some patients show strange behavior with music; more important, though, is the decline in compositional competence, which exceeds the development of a personal style by applying inappropriate or even banal musical material. A violation of musical syntax often seems evident. It is, of course, more evident, when the life of an artist can be followed up. But what about Robert Schumann? Did he relate the delusional thoughts he certainly had several times in his life? Probably not (see Steinberg 1988). During his depressed episodes he stopped performing and composing, as many musicians, professional or amateur, do. Also his late "Waldszenen" do not tell us more than his exciting music can tell. The "Gesänge der Frühe" may show some decline in compositional competence due to the final brain disease Schumann died from.

What always can be heard and easily understood is the emotion a piece of music is played with. The annoying stereotype of hebephrenic mood is noticed just as the ecstatic mood is in which the female patient listened to her own piano playing. We hear whether a player is angry, furious, depressed, or completely unenthusiastic. Mastery in music means understanding the character of a piece and executing it technically in an appropriate manner but also with regard to affective conventions. Of course, affective conventions vary with cultural background and to some extent also according to ones own preferences, but a real deviation from affective conventions and norms occurs only in mental disease. Imitation of the altered, inappropriate affect helps us to understand the madness of a "Wahnsinnsarie," just as happens with real musical delusions.

References

Born G (1982) Mozarts Musiksprache. Kindler, Munich

In der Beeck M (1997) Schizophrenie nach Noten. Psychopathologisches Ausdrucksverhalten am Beispiel des Patienten E. S. Musik Med 3: 29-35

In der Beeck M (1987) Schizophrenic opera - world premiere. Jpn Bull Art Ther 18: 81-85

Kaplan HI, Sadock BJ (1991) Glossary of psychiatry and psychology. Williams and Wilkins, Baltimore

Pethö B (1967) Von der Psychopathologie des musikalischen Schaffens und der Geistesstruktur der Schizophrenen. Confin Psychiatr 10:177-209

Pethö B (1980) Über die Werke eines schizophrenen Komponisten aufgrund einer 13jährigen katamnestischen Untersuchung. Confin Psychiatr 23:250-261

Spoerri E, Glaesemer J (1976) Adolf Wölfli. Baseler Druck- und Verlasanstalt, Basel

Steinberg R (1987) Musikpsychopathologie - Musikalischer Ausdruck und psychische Krankheit. *Habilitationsschrift,* Ludwig Maximilians Universit, Munich

Steinberg R (1988) Robert Schumann in der pathographischen Literatur. Psycho 12:937-949

Streiff P, Keller K (1976) Adolf Wölfli als Musiker. In: Spoerri E, Glaesener J (eds) Katalog der Adolf Wölfli Ausstellung. Druck- und Verlagsanstalt, Basel

Aggression And Autoaggression in the Opera

Walter Poeldinger

La Giaconda

In investigating the problem of suicide in the opera, we encounter an operatic aria entitled "Suicid" in the opera La Giaconda by Amilcare Ponchielli, best known for its fascinating ballet music "The Dance Of The Hours." La Giaconda, a young prostitute, loves the Genuese nobleman Enzio Grimaldo who is also loved by Laura, the wife of a states-man. The statesman attempts to force La Giaconda, who has been brought to him accused of adultery with Enzio, to commit suicide. But Laura exchanges the poison for a sleeping potion. She finds herself in a conflict of conscience, as she herself loves Enzio. In the end she decides to let the two lovers escape and stabs herself.

From the point of view of suicidology, her condition is one of a situative contraction of possibilities, and in her aria "Suicid" we recognize aggression and autoaggression but also the wavering between the wish to die and the wish to live. Finally she decides for death in order to save the lovers. She recognizes that she is now on the last crossroad of her life, and she follows the call of destiny.

In the aria both text and music illustrate her feelings developing from uncertainty to her final decision of suicide. At first she touches a dagger and then drinks a flask of poison. (Singing)...

Gioconda

(Before the table, looks at the dagger, touches it, then picks up the flask of poison.)

"I'll need you! By my own hand I shall do it...
My heart and conscience...they drive me to it!
I can hear voices hauntingly calling:
'By your own choice set an end to it all!'
Laughter and sunlight do I remember...
Where is my Mother? Love lies in embers! I, who
took part in saving my rival, For my own heart I see
no survival!
My day has ended! God grant me pardon When I've
descended into my tomb!

(Looking at the flask.)

This had been meant for Laura. Another victim.
Soon will drink this poison, I myself! Why should I,
When he comes here to join her, remain a witness
To their fond embraces?
Who else can help them to flee to safety? Ah, no!
Tempt me no more! Out of my sight! Advise me,
God given powers of my soul immortal!

(She puts the flask back onto the table.)

Laura lies there... is she breathing... living... lifeless? Who knows? Suppose she were dead...!
Heaven knows how sincerely I've tried to save her!

Yet... if she were dead? An air of odd excitement
Is disturbing my heart! Let's see! I wonder...

(She takes the lantern, goes near the bed, then reconsiders.)

But no! How could I? How could I?
I must not yield to such a thought, not ever!
Yet, were she living... if so, I am alone with Laura...

The darkness hides whatever happens...

(Darkly.)

No one would see me... lagoons will keep their secrets..."

Lucia di Lammermoor

Lucia di Lammermoor decides to marry a nobleman whom she does not love but whom her brother chose for her in order to overcome his financial difficulties. She actually loves Edgar of Ravenswood, who had to leave the country for a while after they had secretly pledged their troth. Lucia´s brother hid letters from Edgar to her and also intercepted her letters to him. The moment she agrees to this marriage, deeply disappointed in her great love, Edgar appears, accuses her of unfaithfulness and throws the ring they had secretly exchanged at her. During the wedding night she stabs her new husband and goes mad.

Donizetti, who ended his own life in madness, actually wrote a textbook of psychiatry in the libretto of his opera. Beginning with the hate between the two families he depicts a brother who is prepared to sacrifice his sister's happiness for the sake of money. We experience an alternating between passionate love and hate. In the hopelessness of her situation she yearns for death and goes mad. In her madness, Lucia projects the loved one, who apparently disappointed and betrayed her, on her unloved husband so as not to see any guilt in him because she did not know of this betrayal. The music expresses this delusive misunderstanding. Thus, the aggression against the man who had apparently treated her so shamefully is unconsciously turned against the person whom she wed and whom she then kills with his own sword. Sigmund Freud would have fully agreed with this part of the libretto because according to his psychoanalytic system this hate is addressed not to her own husband but in this projection also to herself. Lucia does not commit suicide, but she dies in her madness, in that autoaggression first destroys her spirit and finally also her body. In the famous madness aria she sings: (Singing)...

Lucia

"I hear
the sweet sound of his voice! Ah, that voice!
It pierces my heart!
Now I am restored to you;
Edgardo! Ah, my beloved Edgardo!
Yes, I am restored to you!
I have escaped from your enemies...
An icy chill runs through me!
My whole being trembles... My footsteps falter!

Let us sit together by the fountain for a while.

Alas, that gloomy spectre rises to keep us apart! Alas!
Alas, Edgardo! - Ah!
That spectre, that spectre drives us apart!

Let us take refuge before the altar.
The rose petals are scattered! Do you hear that exquisite music? It is the marriage anthem, it is our wedding day!
This is the ceremony! How happy I am!

Edgardo, Edgardo! How happy I am!
My joy is too deep for words!
The incense is lighted - the sacred torches are burning all around us!
Here is the minister... Give me your hand!
Oh, happy day!
At last I am yours and you are mine.
God has united us. "

Tristan and Isolde

After these two examples from Italian opera, which show how differently a conscious or unconscious wish to die can lead to death, I bring a third example from the romantic period of opera. This is the time when Eros was related to Thanatos and the so called "romantic death wish" could be looked at as a particularly ambivalent way of dealing with death.

By desiring death as a relief, all its horror could be substituted by love, and the polar opposites of Eros and Thanatos would lead to a synthesis. A possible, but a dangerous way out. The romantic era reached one of its highest points in Richard Wagner's *Tristan and Isolde*. After Tristan´s death Isolde dies the death of love as Wagner himself called it. In the sense of an autosuggestion she heightens more and more the idea of unity in death to such an extent that she finally dies with the following words: (Singing)...

Isolde

"Can it be that I alone
Hear this wondrous, glorius tone,
Softly stealing,
All revealing,
Mildly glowing,
From him flowing,
Thro' me pouring,
Rising, soaring,
Boldly singing,
Round me ringing?
Brighter growing,
O'er me flowing,
Are they waves
Of tender radiance?
Are they clouds
Of wonderful fragrance?
They are rising
High around me,
Shall I breathe them,
Shall I hear them?
Shall I taste them,
Dive beneath them?
Drown in tide
Of melting sweetness?
In the rapturous swell,
In the turbulent spell,
In the welcoming wave,
Holding all.

I'm sinking,
I'm drowning,
Unaware,
Highest love!"

"I am sinking, I am drowning - unaware - highest love," these are Isolde´s last words. Thus the word "unaware" is already present in the Tristan completed in 1849 in Lucerne before it premiered in Munich. When one listens to Isolde´s "death of love" scene and compares the music with the words, one sees that she brings herself more and more into a state of consciousness that encompasses all senses, so that finally she feels herself drowning or sinking in the universe. With the words "unbewußt, höchste Lust" - unaware - literally "unconscious, highest love!" she dies. One cannot help comparing these words with Friedrich Nietzsche´s "drinking songs": "all desire wants eternity, the deepest, most long lasting eternity!" Therefore just as Richard Wagner´s use of the word "unconscious" in Isolde´s "death of love" predicts a later concept of psychoanalysis, so could Friedrich Nietzsche, with his "Philosophy of looking round the corner" also be considered a forerunner of psychoanalysis.

Wagner also was the first to use the concept of "it" (*es* in German) in his works as the compulsive spontaneous function that he compares with the rational arbitrary one. He also used the word "necessity" to describe the situation in the development from general need to specific human necessity. He had already used the word "it" in a letter to king Ludwig II of Bavaria dated January 26, 1867. He signed this letter "old master of the 'it'." With this description he also had a premonition of Jung´s concept of the shadow and anima when he described Elsa as the unconscious of Lohengrin.

In her famous aria Gioconda considers suicide and also faces the ambivalence between wishing to die and wishing to live. She decides to commit suicide and does so by stabbing herself with a dagger, a manner rare for women. Lucia does not die because she consciously takes her own life, but rather she falls into an episode of madness out of deep despair and deepest hate that makes her become a murderess. She has the hallucination of burning candles and wedding bells in a kind of psychosis of happiness. In it she is united with her beloved and experiences this as a gift of God. Her body is not able to stand the psychotic ecstasy and thus one could explain her death as psychosomatic.

In contrast, Isolde is aware that she is hurrying to her deeply wounded Tristan to save his life as she had done once before. As she approaches Tristan´s corpse she hears a melody within herself, which could be understood as a kind of hallucination. This grows stronger until finally she feels the winds, smells the sea, and asks herself if she should die in this water. At last she experiences herself as an empty vessel united with the universe, and in this drowning and letting go the light of her life is extinguished. This also is a psychosomatic aspect of autoaggression that permits her to die because she cannot live alone any longer.

Dramatization of Madness in the Opera: SALOMÈ, by Richard Strauss

ROSEMARIE EBNER

Introduction

In 1905 Richard Strauss composed *Salomé*, a music drama in one act after Oscar Wilde' s poem (written in 1893 as a dramatic ballad). Wilde mentions it in *De Profundis*: "It contains paraphrases making it to a piece of music." But the two artists were never able to discuss a Salomé project, as Wilde had already died 3 years before Strauss ever saw this play.

The drama is rather problematic; the story takes place at a turning point in history and revolves around the confrontation of two civilizations. Strauss himself said about the rather difficult ending of his composition: "...it is unfathomable, and even I cannot explain it exactly, otherwise I'd not be an artist but a journalist!" To save time I discuss only a few selected examples out of the fourth scene of Salomé, the last scene of the piece in which the climax of the drama occurs. This scene consists of two parts, interrupted by Salomé's famous dance of the seven veils.

Scene Four, Part I

From Herod's monologue it can be seen that Herod's specific character breeds Salomé's destructive leanings. At this time he represents an extremely unstable person, superstitious and full of fear. Strauss represents this aimlessness musically by the wholetone scale, which does not define its modality as major or minor and has no leading tone (Fig. 1).

Fig. 1. Wholetone scale

Symbols

> Ah, the moon looketh strange to-night!...Much like a drunken woman that is reeling through the clouds...

In the tetrarch's vision this woman in the clouds symbolizes the princess Salomé who is yearn to kiss the prophet's mouth.
The two first sentences of the drama are

> "How fair is the Princess Salomé to-night!" and "See the moon, oh look, how strange the moon seems!"

and from another moon vision which associates Salomé's beauty and the moon's sallowness.

Beauty	-	sensualism
Sallowness	-	death

These contrasts foreshadow Salomé's fate. In his music Strauss associates these visions closely, referring to the first moon vision harmonically (C.6.3 of the score)[1] by motife 33 (Fig. 2) with the same seventh chord:

Fig. 2. Motif 33 moonvision. Motif of harm

and the fact that in both moments the sight of the moon evokes uneasiness for those present.
Other symbols used by Herod are:

> "In blood I have slipped, it is an ill omen." or: "There is a wind - and in the air I hear something like the beating of vast wings" (a well-known symbol for death).

Herod's Erotic Relationship to Salomé

In this fourth scene Herod courts Salomé unsuccessfully, probably not necessarily because she dislikes him, but rather because she feels spiritually empty after being

[1] C. means cipher, counted from each cipher in the score or in the vocal score. For example, C.1,5 means measure 5 after cipher 1.

so roughly rejected by Jokanaan; however, the prophet remains the object of her desire. After twice refusing the tetrarch's request to dance for him, Salomé fatally induces him to promise her the fulfillment of any wish she would make.

Strauss expresses this by representing the mutual feeling musically in the motifs 40 and 41 (Fig. 3a, b).

Fig. 3. a (M 40) Herod's promise I, C.226,7/227,1

Fig. 3. b (M.41) Herod's promise II,C.228,1

The decisive dialogue is:

"Will you indeed give me that so ever I shall ask, Tetrarch?"
"All, all what so ever thou shall ask of me..."
"You swear it, Tetrarch?"
"I swear it, Salomé."
"By what will you swear it, Tetrarch?"
"By my life, and by my crown, and by my gods."
"Remember that oath you've taken, Tetrarch."

With the critical word "ALL..." an ecstatic flute solo starts (C.227,6), which derives directly from the Salomé motive (Motif 1). A long trill unifies these sections a "promise, oath and confirmation," and then Salomé says: "I'm ready now, My lord," and dances.

Dance of the Seven Veils

The musicians begin a wild dance "molto presto e furioso," while Salomé stands motionless. When she rises and makes a sign to the musicians, they subdue the wild rhythm instantly and begin to play in a soft and swaying tune.

This dance not only refers musically to the fin de siècle or the overexcited atmosphere of the oriental court, but it also symbolizes the underlying thought of extreme glorification of bacchnalian effusiveness, developed to a celebration of the Dionysian spirit. Similar to Brünnhild's final monologue in the Wagner Trilogy the dramatic form of the music has a typical ritualistic character. The musical coherence is based on themes already known (remembrance), on the one hand, and the anticipation of the sequence, on the other. After the "molto presto" finale of her dance Salomé remains for an instant in a visionary attitude near the cistern where Jokanaan is kept prisoner - then she throws herself at Herod's feet, who says: "I'll pay thee royally, my princess. I'll give thee what ever thy soul desireth...Speak!"

The composer pictures Herod's enthusiastic excitement in the following sixteenth and eighty-note rhythms of a variation of motif 33 (Fig. 2) which is now much more animated than it was originally and follows the character of the dance.

Herod's Conflict with Salomé

The tetrarch realizes the desperate situation he is in and therefore tries to influence Salomé; however, he cannot suggest but unsatisfactory offers. In the 17 sections of the opera they try to resolve their contradictory opinion. Strauss begins with a variation of the second "motife of harm." But Salomé demands uncompromisingly "the Head of Jokanaan." Thus her statements make Herod more and more desperate and his propositions useless. The music expresses this difference by changing the thematic material and the tempo. While Herod's very excited sections are marked "accelerando," "presto" or even "alla breve," Strauss wants the Salomé parts to be played "*doppelt so langsam,*" that is, twice as slowly. When Herod realizes the senselessness of any further discussion the trombones play the wholetone scale again. While the tetrarch is frightened by a nightmare, the music changes from polytonality unequivocally to the key of D minor. With the motif symbolizing the "desire of destruction" (Fig. 4).

Fig. 4. Motif. Salomé's desire of destruction

Strauss inserts the fixed point of the conflict. The composer orchestrates the 17th and last section expressively with strings and brass. Herod's stumbling and slipping is represented by motif 58 (Fig. 5).

Fig.5. Motif 58. Herod's motive of stumbling and slipping (C.153.5)

After a final confrontation which Strauss composes bitonally in D minor and E flat minor, he expresses the incompatibility of the opinions musically in the sections 8,

10, and 12 by confronting the two keys in tritonal relationship. Trills, intervalls of a third, and tone repetitions additionally underline Salomé's inflexible attitude. Besides different musical figures which are not used again, there are basic motifs which recur again and again, such as motif 43, motifs 41 and 46, the wholetone scale, and others (see Fig. 6).

Fig. 6. a Motif 43 Herod's third promise C.251,3.

Fig. 6. b Motif 46. Herod's fear

With the exception of motif 31, none of the motifes continues to be used until the end of the dialogue. The different sections of this scene are both joined and separated by Salomé's repeated demand for "Jokanaans head".

Scene 4, Part II

Salomé is enjoying her triumph while her mother takes the "ring of death" off the hand of the tetrarch and gives it to the first soldier; he delivers it to the excecutioner, who descends into the cistern. Strauss calls for a tremolo on E flat in the double basses, which, according to the score, has to sound like "the moaning of a woman." The scéne climaxes with motif 19c (the motif of eroticism) (Fig. 7).

Fig. 7. Motif 19c. Salomé's motif of eroticism, here at a fast tempo and motif 31 (desire of destruction) played in an unrestrained manner, thus expressing the ambiguous character of this orgy.

In this music the composer shows the unbridled wildness of the princess and reminds the listener of the fact that all this ecstatic activity aims at killing and destroying for the sake of raising one's own pleasure of living.

In the final monologue the individual motif recurs in constantly changing musical relationships, which contributes to the musical effectiveness of this section.

Conclusion

The ending of Wilde's drama allows for several different interpretations. However, none can explain musically the Strauss opera satisfactorily. Though for Strauss, too, Salomé was first predestined to perish in the final monologue musically the two sides of Dionysian spirit - the one full of sensuous interest and the other allied to the thought of death - are connected.

But Strauss is inspired by a utopian idea and composes the words addressed to Jokanaan "If thou hadst looked at me, thou would have loved me" in "Salomè's key" of C sharp major. The prophet's motive (motif 5) is also transposed here to C sharp major and combined with the Salomé motif I. The composer then continues his vision with the passion motive (motif 21) and by the princess' significant statement: "And the great mystery of love is greater than the mystery of death..." As very often in the nineteenth century, love and death are closely related here, too; this comparison enables Strauss to reinterpret the Salomé material and thus to effect the connection of Salomé's "desire to kiss" and its fulfillment musically by repeating motif 42 of the end of the dance, where it symbolizes her "Dionysian vision of fulfillment." Strauss interrupts the sequence of the now regularly recurring "kiss chord" only to express again and again the motif of "vision of fulfillment."

Thereafter, the usual harmonic cadence of fourth-sixth chord, dominant, the tonic follows in C sharp major (360.5). Motif 52 remains predominant in this whole section until the ending (C.361.1). The chord in C.360,6 has given rise to different interpretations, but expecially the following seems interesting: Strauss raises the "death chord" half a step and it turns into the well-known "kiss chord." So this "death chord" - "kiss chord" combination symbolizes two different sides of the mystery of life, as a transition from horror to pleasure, another perspective well-known from the *Birth of Tragedy* by Nietzsche. This philosopher describes the Dionysian attitude as a "destruction of the ego experienced with delight, being at the same time a happiness to perish and a chance of transition."

It would be difficult to find a better musical representation of this paradox than the one Strauss has created in Salomé. Whether he composed it intuitively or fully consciously, the composer had accepted Nietzsche's thought of the "will for eternity" and this might have found a creative realization in the ending of his Salomè.

References

Natan A (1963) Richard Strauss, Opern. Basilius, Basel
Newman E (1970) Richard Strauss, testament of music. Lande, London (Original 1970)
Schuh W (ed) (1953) Recollections and reflections. Boosey and Hawkes, London
Strauss R: SALOME-op.54, music drama in one act after Oscar Wilde's poem, transl by Hedwig Lachmann - score Berlin 1933, Ad. Fürstner, Boosey and Hawkes, London (No 9124)

Can Strong Experiences of Music Have Therapeutic Implications?

Alf Gabrielsson and Siv Lindström

Introduction

Music in the treatment of physical or mental illness is documented since ancient times, and music and medicine form "a partnership in history" (Pratt and Jones 1987). In contemporary music therapy, music is used in a variety of ways, depending on the problems and the purpose of the treatment (see, for instance, in Bruscia 1987, Frohne-Hagemann 1990, Spintge and Droh 1987, Strobel and Huppmann 1991, and Unkefer 1990). Without directly referring to therapy, people often talk about the importance of music in their lives, not least during personal crises and describe how music can change their state of mind.

Strong Experiences of Music

Methods

In an ongoing research project on strong experiences of music (SEM) (Gabrielsson 1989, 1991; Gabrielsson and Lindström, in preparation) we found several interesting examples of potentially therapeutic implications of music. In order to provide a background for these examples we first briefly describe the SEM project. The aim is to get a description and classification of strong experiences of music and to investigate which factors - in the music, the person, and the situation - contribute to such experiences. We asked subjects to describe the strongest experience of music that they have ever had and to answer various questions concerning the music, the situation, their physical and mental state, and music preferences. They were also given a large set of statements concerning strong experiences of music and were asked to indicate how well each statement agreed with their own experience, using a scale from 10 (perfect agreement) to 0 (no agreement at all).

About 800 persons have participated, more women (about 62 % of all participants) than men (about 38 %) and musicians as well as nonmusicians. Their age ranges from 13 to 91 years, and their musical preferences are spread across many genres - classical music, pop/rock, jazz, folk music, and various types of popular music.

The descriptions given by the persons in interviews or by means of questionnaires are subjected to content analysis in order to collect and systematize all relevant phenomena and aspects of SEM. This is by far the most demanding part of the analysis. The quantitative data from the agreement scales are analyzed by conventional descriptive statistics and factor analysis. Both the qualitative and the quantitative analysis indicate a very large number of aspects of SEM. A brief classification of these may be made as follows.

Aspects of Strong Experiences of Music

Strong experiences of music may involve:

- *Physical responses*, such as shivers, gooseflesh, changes in breathing and heart rate, various overt movements, tears in the eyes, weeping.
- *Quasi-physical responses*, such as feeling weightless, being carried away by the music.
- *Perceptual aspects*, related to various aspects of the music (timbre, rhythm, melody etc), visual impressions of the performers, and tactile sensations.
- *Cognitive transformations*, such as total concentration on the music, lost awareness of time, space, and one's own body, abandoning of analytical listening, getting images/fantasies.
- *Cognitive - emotional aspects*, such as experience of direct communication without any barriers, clarification of own thoughts and feelings, and a sense of perfection.
- *Emotional responses*, such as joy, happiness, excitement, euphoria, love, warmth, peace; in some few instances negative feelings as grief, anxiety, pain, and anger.
- *Transcendental and existential aspects*, such as experience of wholeness, cosmos, presence of God, reflections on the meaning of reality and existence.

The examples for each category given here are but a few out of a multitude in our material. A detailed account will appear in Gabrielsson and Lindström (in preparation).

Therapeutic Implications

Individual Reports

In the following we focus on potentially therapeutic implications of SEM. A number of individual reports are given, followed by an attempt at classification.

Person A

About 20 years ago I met with an event that gave me a deep personal shock. This meant that I could not get any contact with people and the life around me. I felt as if I lived in a bell glass with absolute isolation from life. The only thing that reached into myself was music. I was lying in a dark room and listened to music from a gramophone. The music could break through the walls and trickle into me, and I believe that it was the music that meant that I, after a month or two, could begin groping my way out into life again.

Person B

My strongest experience of music ever was in March 19.., when I had just got a recording of *Rachmaninov's Piano Concerto No. 2 in C minor.* I brought it together with a simple, battery-driven plastic gramophone, when on my own demand I was taken into a psychiatric clinic during a severe crisis. Since I had a private room, I was able to play this fantastic music from early morning to late night. But I always had the door open in order that as many people as possible would be able to get uplifted and feel all the indescribable things that I felt. But people's lukewarm response surprised me greatly.

I was in a deep depression, but the more I listened to the music, the more my thoughts became lighter. Everything was so sad and the music was so sad to me, but it made me feel more and more that life was coming back. It became light around me quite obviously. It was as if the sun was shining straight into my room, straight into my heart. This process went on for 4 weeks. Every bar, every note in all of this fantastic piano concerto was so infinitely dear to me, so indescribably beautiful and varying. I could hardly understand that something so heavenly and beautiful existed. It then took 15 whole years before I managed to listen to this music again, so strong were the associations.

Person C

Person C talked about listening to a record with Gregorian chant performed by a choir from the Solesmes monastery:

I played this music for a period of about 10 days in November 19... I was in my home, alone, after an operation with a stay in hospital for 10 days. In the hospital I got morphine against intestinal cramps. I took no medicine against the pain after coming home. I had bought the music...1 week before the operation (that was planned), as if I understood (not consciously) that I was soon going to need this music... The strongest experience of the music was on the first day. My description in the following concerns experiences that recurred at every attack of pain when I put on the music. I listened for several hours every day. The attacks lasted 20-30 min and turned into a dull pain that was there all the time and disappeared wholly after about 2 weeks.

Then comes an explanation of the relieving effect of the music:

The Gregorian chant is humming and means consolation. I go down into the ancient church, down into my own archaic rooms, the rooms where somebody was humming. It is a clear picture. Ingrid wore long white aprons. When she died, I was 1 year and 10 months old and was just pining away. The whooping-cough I got afterwards lasted for almost a year. It was a lengthy and life-decisive experience. All the time I remembered Ingrid's humming and in that way I had the consolation within myself. I revived this in November 19... I needed to be alone and had difficulties talking with anyone. I felt that I could only go through this quite on my own. I played Gregorian music and pulled myself through the pain. The external (the music) and the internal (Ingrid) humming met.

Person D

Before the birth of my first child there were careful preparations in many ways. Among other things, these included preparing for a possibly lengthy delivery. In order to fill the time, I put together a tape especially for this occasion, later called "the delivery tape", comprising music that I quite simply like. Significant music of different kinds, "pearls" for me... However, this tape proved to be far more important than I then believed... When the labor started and it began to get more painful than bearable, my husband took out this tape that I by then had completely forgotten. I remember that it in that situation seemed irrelevant and unnecessary to listen to music, since the only thing I experienced was pain. Despite my weak protests, the tape was put on and the small, unfamiliar delivery room in the... hospital was filled with music. My own music.

Practically immediately, and in an indescribable and quite new way, my music filled me with power, courage, and safety in such a strong way that no later experience of music has been able to measure itself with this one. The pain of the labor faded; it was there but subordinate to the music, quite manageable and not painful any longer. All other medical pain reduction was felt quite unnecessary and was rapidly interrupted... That in this unfamiliar and quite new situation get my own music changed my whole situation. A "pain reduction" of the most efficient kind, both mental and physical. Another one of music´s power and possibilies, through whose help my daughter´s birth became a very positive event containing great presence, in me, concerning what actually happened during the delivery and great happiness.

Person E

I got the chance to borrow a cassette with this music [Händel´s "Concerti grossi," Op. 6] from a friend and now play it on a walkman. I am lying on my bed [in hotel S...], looking at the moon and the sea. I notice my usual irritation over the "strutting" character of classical music. I feel it as dry, artificial, and intellectual. I notice that the music does not seize me in the same way as good rock music does. I think that if the music doesn't seize me, can I seize the music? I feel that there is a resistance in the center of my body. It is as if I do not want to follow or let me be led by the music. I think that I ought to be able to follow the music with all parts of myself. That is, if I give up the resistance that I feel, maybe I can understand what it is that captures other people in this music. I try to make myself receptive and soft internally...

Here we come to No. 6 [Musette Larghetto], and something loosens its grip within me. The beauty and the depth I feel is boundless. It is as if places within me that have grown stiff since many, many years, now soften and take part in the music. But it does not occur passively, there is a definite quality of really following. There is total freedom just now. Now it lives in a place within me, which used to be dead. The place that is living is the place that gives depth and the contact with life, the zest itself and the taste. I get up from my bed and stand looking out at the sea. I realize that normally I live enclosed and without this deeper contact with reality. But here it appears fresh and newborn and all the time new. As it probably was when I was a child. The music is about forgiveness. An old burden is relieved , and I realize that all that I normally wish is here. And it has always been there, right in front of my eyes.

Person F

Last Christmas...I was in a phase in which I could not cry, it felt as if I had no contact with my feelings - they never came up to the surface. Then I got to hear "Mary's Tune" performed by Annika Skoglund on a record with a musical written by Tomas Boström. The musical is about different women in the Bible, and the song that Annika Skoglund sang was about Mary Magdalen´s pain when she was faced with the fact that Jesus was dead and not together with her any longer. "How will I be able to live without him"? she sings. Even though I cannot fully accept Annika Skoglund's way of singing, the words of the tune went straight into my heart, and the hard armor around my feelings opened up and I could both cry and laugh again... In Mary's cry and pain I saw the cry and pain of all women having lost their beloved (husband or children). It is a tune that could be used in a widow's mourning work after a long and happy marriage. Perhaps it was primarily that that I realized - that the tune was not only an unusually good tune but that it served a function. That one through this tune could get help to live out one's own sorrow, anger and pain. After having heard this tune, I walked round during the Christmas holidays and was incredibly sensitive and easily moved and was also tremendously happy about being able to be like that.

Person G

I had my strongest experience of music in October 19... Our best friends' 19-year- old son was to be buried after having died in a car accident. I, my family plus about 400 more people were "listeners" attending the funeral ceremony. I felt an indescribable grief, grief, grief... An overcast day, with low clouds and the rain hanging in the air. It was heavy in many ways. When the funeral ceremony had lasted for about half an hour, a flute began to play (perhaps the organ too, I am not certain, but it is the flute that I remember), and after some bars the sun began to shine through the altar windows of different colors, and the whole church got light and one "lifts" in a way, and it felt as some kind of surprise, relief, consolation. When these different feelings appeared, I did not know that the piece being played is called "The Fog Is Lifting" by Carl Nielsen, but I saw it after a while in the program. A very remarkable experience. An event that I have thought much about later. Perhaps as an attempt at consolation for myself?!... In the beginning of 19... I experienced exactly the same feelings, grief - surprise - relief, when the same piece was played on the radio. After only two to three tones I began to cry and had to stop at the side of the road (it was in the car), but the grief rapidly disappeared and everything felt light, consoling.

Person H

This happened during autumn 19..., and I was in my room at home. I was sad and depressed and felt that the only thing that would be able to help me was music. So I pulled down the curtains, switched off all lights and put on one of my favorite records. It is "The Trio Album" with my absolute favorite singer Dolly Parton and the two country singers Linda Ronstadt and Emmy-Lou Harris... I was sitting with my eyes closed on the floor and felt the music, at the same time as I was singing Dolly's part, the highest. We were singing together in such harmony that I have never experienced anything similar earlier. It felt as if I was soaring and was lifted higher and higher up by the wings of the music! All problems disappeared and my whole body was dissolved into tones. (I want to add that if I am able to choose, it is like this I want to die some time in the future, because I have never experienced anything so wonderful!) When the record was finished, it felt as if I had slept for 24 h at a stretch. I felt so thoroughly rested! This meant that I use the record when I am depressed (other times too, but especially then). It provides some internal cleansing but only in combination with darkness and loud volume (but not so loud that my own voice is not heard, because I have to sing myself if it is going to help).

Discussing the effects and use of music, another person said:

Person I

It may be a common pop/rock tune. Then I may feel: Oh, this music is made for me. I feel myself selected, acquainted. A feeling of community. There are people feeling like I do. They express themselves like this, and if I had done it, I would have done it in the same way. It is funny with music... it helps me in my life.

Sometimes I want to confirm or strengthen a feeling. For instance, if I feel a little sad or melancholic, then I want to play such music. Now I know what tunes contain this, touch that feeling. Or if I come home and am super-happy, I take an inspiring tune that makes everything even happier. I can also do the reverse - change my mood with the help of music. I may be getting a little sad, then I can decide that: No, now I must put on "this or that" to make me happier and get some confidence again. And it can really work like a snap of the finger. I get quite incredibly affected by music - one can certainly say that. Well, it is good; it is like therapy. It is a good trick when you learn how it works. It is nice. It is a resource.

Now listening to flamenco music performed by a female guitar player:

Person I

It is something that touches you very deeply... it feels as though it has to do with the existence itself, the life. It is such an extremely strong experience that you are here and now, nothing else matters, you just exist. One gets such an extremely strong feeling of life, of living. There is everything, life, death, existing as a human being - all pain, grief, passion, joy. It is as if everything is compressed into these tones. It is only she and me who exist in this room and we both know what it is all about, it is about living. There is something vibrating that reaches a string within yourself. Otherwise it is just lying there but now something makes it wake up. I just simply get a strong experience of myself, my ego. I live. I am. I exist.

Classification

These reports suggest different kinds of therapeutic implications related to SEM. They may be briefly classified as follows:

1. Music may provide relief from physical pain (person C after an operation, person D in connection with childbirth).
2. Music may alleviate deep grief and provide consolation and new hope (person G attending a funeral ceremony).
3. Music may help in overcoming temporary depression (persons H and I) and even severe, long lasting depression (persons A and B).
4. Music may release barriers or defenses, often unconscious, giving contact with hidden thoughts and feelings and providing feeling of openness and freedom (persons E and F).
5. Music may provide new insights concerning one's way of living, relations to other people and to reality in general (person E).
6. Music may make oneself feel acquainted with, and confirmed by, other people, thereby increasing one's self-confidence (persons H and I).
7. Music may be deliberately used in order to confirm, strengthen, or change one's state of mind (persons H and I). In this sense one may learn to develop a therapeutic repertory of music.
8. Music may create a strong feeling of one's ego, the conditions of human life and existence (person I).

Other reports not included here tell about the possibility of music to provide experiences far beyond common verbal and intellectual comprehension, visions of other realities, "glimpses of other kinds of existence", and deep spiritual or religious experiences. Some people declared that the experience meant a total change and re-orientation of their life regarding attitudes and value systems.

Discussion

The examples and categories given here are by necessity limited. Furthermore, the categories are unavoidably more or less overlapping, since many of the implications go together. Examples of other implications were given in Gabrielsson and Lindström (1992) and will also appear in forthcoming publications.

In the reports above, the music was more or less deliberately chosen beforehand by some individuals (persons B, C, D, H, and at least partly by persons E and I), whereas for other individuals the music just happened to appear in a certain situation (persons F and G, maybe person A). The therapeutic effects thus seem to be planned or expected in some cases but just happen unexpectedly in other cases. (However, the performance of Carl Nielsen's "The Fog Is Lifting" in a funeral ceremony may reflect a therapeutic intention from the person responsible for the selection of this music.)

A detailed discussion of these questions in relation to various music therapy techniques is beyond the scope of this paper. As examples of techniques somehow related to SEM we may just briefly refer to the "Guided Imagery and Music" approach by Bonny and Savary (1990) and various techniques in receptive music therapy as discussed by Schwabe (1978). In these techniques music is used as a way of eliciting experiences and responses in the client related to his/her problems, which then may be accessible for further treatment by verbal discussion and interpretation. Musical and verbal means are thus integrated. In the present reports we find examples of therapeutic implications of music without further verbal interpretation.

From the evidence presented above we think that the question posed in this paper - Can strong experiences of music have therapeutic implications? - can be positively answered. More generally, we think that our SEM project convincingly demonstrates the power of music to affect individuals and groups in ways that few other means are capable of doing.

Acknowledgments. We thank David J. Hargreaves for supervising the English. This research was supported by The Bank of Sweden Tercentenary Foundation.

References

Bonny HL, Savary LM (1990) Music and your mind, 2nd edn. Station Hill, Barrytown
Bruscia KE (1987) Improvisational models of music therapy. Thomas, Springfield

Frohne-Hagemann I (ed)(1990) Musik und Gestalt. Klinische Musiktherapie als integrative Psychotherapie. Junferman, Paderborn

Gabrielsson A (1989) Intense emotional experiences of music. First International Conference on Music Perception and Cognition, Oct, 17-19, Kyoto, pp 371-376

Gabrielsson A (1991) Experiencing music. Can J Res Music Educ 33:21-26

Gabrielsson A, Lindström S (1992) Esperienze musicali intense e loro implicazioni nello sviluppo personale. In: Pritoni LR (ed) La musicoterapia. Implicazioni cliniche e psicopedagogiche. del Cerro, Pisa, pp 11-18

Pratt RR, Jones RW (1987) Music and medicine: a partnership in history. In: Spintge R, Droh R (eds) Music in medicine. Springer, Berlin Heidelberg New York, pp 377-388

Schwabe C (1978) Methodik der Musiktherapie und deren theoretische Grundlagen. Barth, Leipzig

Spintge R, Droh R (eds) (1987) Music in medicine. Springer, Berlin, Heidelberg, New York

Strobel W, Huppmann G (eds) (1991) Musiktherapie. Grundlagen, Formen, Möglichkeiten, 2nd edn. Hogrefe, Göttingen

Unkefer RF (ed) (1990) Music therapy in the treatment of adults with mental disorders. Schirmer, New York

IV. Brain Imaging and Music

EEG in Music Psychological Studies

Hellmuth Petsche, Peter Richter and Oliver Filz

One of the main aims of the present paper is to comment upon why the EEG seems to be such a promising tool to glimpse - generally speaking - into the fabrics of thinking, the "enchanting loom" as Sherrington formulated it so poetically and so soundly. In addition to that, two special issues concerning music and EEG are briefly discussed: hemispheric lateralization and musical expectancy. Further topics concerning the processing of music by the brain, among these the question of the consistency of our findings are the subject of another paper (Richter et al., this volume).

When the first author started studying epileptic seizures decades ago he never would have thought that the most conspicuous EEG phenomenon in seizures - the fact that often far-extended regions of the brain seem to oscillate fairly synchronously - might one day have such a bearing on the elucidation of cognitive processes. As it has turned out, cognitive processes are characterized in the EEG by different brain regions becoming electrically more closely or less closely coupled, thus indicating closer or looser functional connectedness, which probably hints at different degrees of information flow between these two areas. As has been demonstrated in the past decade, this holds true for all kinds of mental acts, from listening to music to reading (Petsche et al. 1988), from language interpreting to playing chess (Petsche et al. 1993b): all induce topa graphically multifarious patterns of electric coupling and decoupling, reflected by topic syn- and desynchronizations of the EEG, similar to those found in epileptic seizures, but by far weaker and not visible to the naked eye.

When studying epileptic seizure patterns a proper method of determining synchronization quantitatively was urgently needed. Such a method had been developed for seizures and was adapted to our special purposes. We were lucky enough to adapt it, years later, to the EEG of mental acts. We are referring to coherence, a by-product of Fourier analysis, which may yield hints at the strength of coupling between electric oscillations in different brain areas with respect to frequency. Coherence is a measure of cooperativity: our experience has taught us that, when any reflections of thinking processes in the on-going EEG are to be studied, it is of greatest importance to examine spatiotemporal cooperativities in the system "brain."

To better understand how cooperativity may come into being in the brain, a short side trip into neuroanatomy seems worth while. For most neuroscientists the cortex is but an accumulation of an enormous number of nerve cells of different shapes, which usually are arranged in six layers. This conception goes back to

Golgi preparations in which mainly randomly distributed nerve cells are found with dendritic trees and a few axonal ramifications. However, this technique demonstrates only a tiny aspect of the morphological features of the cortex and is liable to create a mental concept of cortical structure wich hinders the exploration of brain functions if brain structure is the starting point: Golgi pictures are indeed often conceived of as a kind of blueprint of the cortex: these blueprints, however, are erroneous since they do not demontrate the most important features of the tissue structure. In Golgi sections only about one thousandth of the actually existing neurons are stained, their selection being due to chance. Therefore, this kind of preparation can evoke only a very rough idea of cortical structure. Unfortunately, the same holds true for all the other numerous methods invented for the detection of cerebral circuitry. Each of them makes visible only a few tiny aspects of cortical structure.

Recently, however, another way of exploring the structure of the brain opened up. Since all attempts to reveal cortical circuitry by bottom-up strategies seemed doomed to failure, statistical approaches to studying the structure of the cortex were attempted. In 1991 Braitenberg and Schüz published a book with the title *Anatomy of the Cortex, Statistics and Geometry*. Braitenberg's life-long experience with counting cell and fiber densities led him to conclusions which shed some light on the purpose of the cortex and thus may help in understanding cortical functions and, with that, our own results too. We quote only a few of his results, those which contribute considerably to an understanding of brain function in general and of our findings in particular.

One of Braitenberg's major conclusions is that no neuron of the cortex is any farther apart than 2 - 3 synapses from another neuron. This means, first, that the cortex has the prerequisites for an extensive information exchange and, secondly, that, on average, almost every neuron "knows" of the activity of any other, at the worst via only a few interconnected neurons. Therefore, every cortical region is informed in some way about the events taking place in the rest of the cortex after a delay of a few synapses only.

Another salient feature momentous for the understanding of cortical function is that every cortical pyramidal cell contacts with almost as many other neurons as it gives off synapses (this is on the average 7000 - 8000 per neuron); no more than about 1/1000 of the synapses found within the territory of one pyramidal cell (as defined by the horizontal extension of its dendritic tree) originates from this cell. The rest of synapses in this territory stems from other cells or regions of the cortex, respectively.

A further conclusion is: the dendritic "clouds" of one neuron, i.e., the area of its dendritic ramifications, can be completely included in those of others, and the axonal "clouds" of one neuron overlap with the dendritic ones of many others.

Therefore, the main function of the cortex seems to be to ensure highest possible convergence and divergence of signals and, thus, to guarantee exchange of information as extensively and intensively as possible. These conclusions are further supported by the observation that the majority of the fibers in the cortex are corticocortical and intracortical fibers: only about 1% of cortical fibers connect the cortex with the thalamus, the basal ganglia, and the spinal cord.

One more point has to be added which may help in understanding the significance of cortical structure for the EEG. The studies of Nunez (1981) on the nature of the electric fields of the brain brought some evidence that the intra- and corticocortical fiber systems are of great importance for the emergence of its electric oscillations, the EEG. This statement underlines the necessity of studying the electric relationships between different brain areas to elucidate any events underlying mental processes.

These prerequisites seem to justifiy focusing on the electric relationships between different areas of the head more than on the absolute size of potentials and their distribution. For our purposes study of the electric relationships also turned out to be more effective. In order to elucidate these relationships, a method based upon Fourier analysis has been developed; it supplies parameters for both aspects of electric activity, relationships and amplitude.

This method is briefly as follows (Table 1). It is a top-down method and as such adequate for analyzing the EEG, which is known to be a mass phenomenon. In principle, EEG sections of at least 1 min during the performance of a mental task are compared with long-lasting EEG sections at rest. Because the EEG is not a stationary process, it is advisable to use longer periods of recording; 1 min is the minimum. The recommendations for minimal duration for such experiments are between 20 s (Möcks et al. 1984) and 1 min (Salinsky et al. 1991). EEG is recorded by means of the clinical 10/20 system with 19 electrodes. Four parameters of the EEG are taken into account: the position of the electrodes, the frequency band, whereby the beta band between 13 and 32 Hz is deliberately subdivided into three bands, beta 1 through 3, and, finally, absolute power or amplitude, respectively, and coherence. Significant changes of these two parameters between task and EEG at rest or between groups of individuals with different charcteristics are evaluated.

As already mentioned, most attention is paid to the parameter coherence, which provides information about the correlation between two signals with respect to frequency. This paper exclusively deals with coherence since most emphasis is put on the idea of the brain as an information processing network; thus, the primary question is, which regions cooperate with one another rather than which areas are most active.

Coherence provides data on the relationships between the electric oscillations recorded from two locations on the skull. In case of no correlation whatsoever, coherence is "0," in case of synchronous activities, coherence is "1" within the frequency range studied. Between these two limits, however, coherence can take every possible value. More casually expressed, coherence can be taken as a parameter measuring the degree of similarity of two EEG signals or of the electric kinship of two brain regions in a certain frequency range.

For the time being not so much care is taken to ascertain the absolute size of coherence, which is different in different brain regions. This method does not consider the phase shift either, but rather only significant increases and decreases

Table 1. Prerequisites for the study of thinking processes by means of the EEG (probability mapping)

"Top-down" rather than "bottom-up" - strategy
EEG sections of at least 1 min
Recording from the entire skull ("10/20-system")
Reduction of the EEG to four parameters:

Recording site (19)
Frequency band (5, between 4 and 32 Hz)
Absolute power
Coherence

Computation of significant changes

Task vs. EEG at rest
(individuals or groups of individuals)

Computation of significant differences

Individual vs. individual
or group "A" or group "B"
(at rest or during aspecific task)

Representation of the results in schematic brain maps

---

with respect to the averaged EEG at rest. As for the interpretation of coherence, at present one may hypothesize that increases may be caused by the activation of more, and perhaps more advantageous, neuronal connections between two cortical regions, whereas when coherence decreases, these connections may be inhibited or the processes taking place during idle thinking may be guided into other neuronal channels more important for the activity in question at the moment. Although this admittedly simple interpretation can only be taken as a working hypothesis for the moment, it has proved helpful in explaining several results.

Next, the representation of significant coherence changes caused by mental processes has to be demonstrated. By using 19 electrodes, a total of 171 coherence increases and decreases can be estimated for each task and each of the five frequency bands. (Fig.1) In order to graphically represent such a great number of possible significant changes of coherence graphically it has turned out to be advantageous to arrange the 19 electrodes along the circumference of a circle, the electrodes of the left hemisphere to the left, and those of the right one to the right. The three electrodes of the midline are on top. In such a graph, straight lines between the respective electrodes indicate significant changes of coherence between these two sites; full lines are for increases, broken lines for decreases. This is made for every frequency band. Moreover, different probabilities of the coherence changes may be entered in this scheme by varying the strength of the lines.

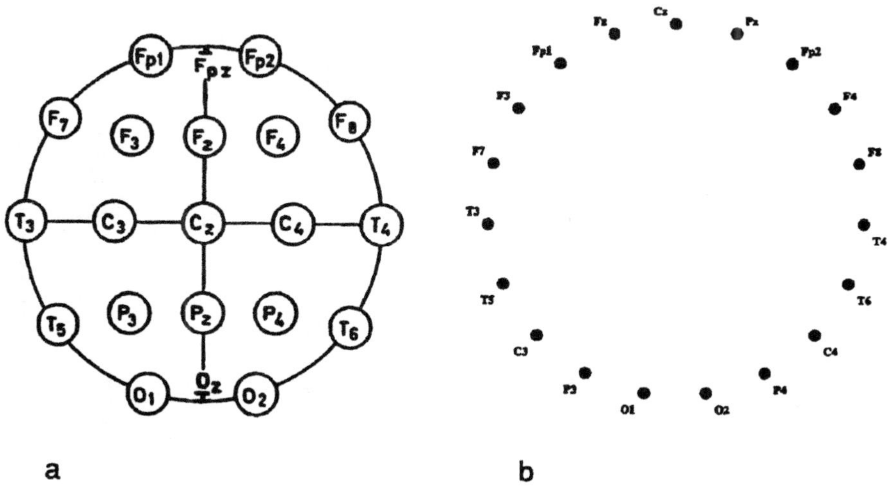

Fig.1. a 10/20 system of electrode placement. b Arrangement of electrodes to enter significant task-related changes with respect to the EEG at rest.

Such a way of representing coherence values has the additional advantage that, for each frequency band, the number of intrahemispheric coherence changes for each side and the number of interhemispheric changes can easily be seen, a condition required for answering the question of laterality in music processing.

An example of this presentation is seen in Fig.2. The coherence patterns were obtained from a 29-year-old male subject who was listening (for 5 min to each piece) to Bach (Prelude and Fugue in a minor, Well-Tempered Clavichord I), Schönberg (op.33, pieces for piano), Beethoven (Sonata op.110, first movement) and to a jazz piece for piano (Aldo Romano: "Amalgame"). In addition, the subject was listening, for the same periods of time, to a German and an English text, and, finally, while mentally performing a simple arithmetic task (for 1 min). The EEG at rest and during all tasks was performed with eyes closed.

The figure presents the significant increases of coherence with respect to the averaged EEG at rest during these seven tasks, represented for each frequency band, one on top of the other. The degree of probability is indicated by the thickness of lines, the thickest lines representing the highest probability ($p =< 0.01$), the thinnest the lowest one ($p =< 0.05$).

The first fact to be noticed is that, during the 5 min while these tasks were being performed, coherence with respect to the EEG at rest became significantly increased between many regions of both hemispheres. Secondly, the two hemispheres were not equally involved and the patterns differ from one frequency band to the other. And finally, the patterns for the different tasks are completely different.

At this point the question may arise as to how far these differences may be due

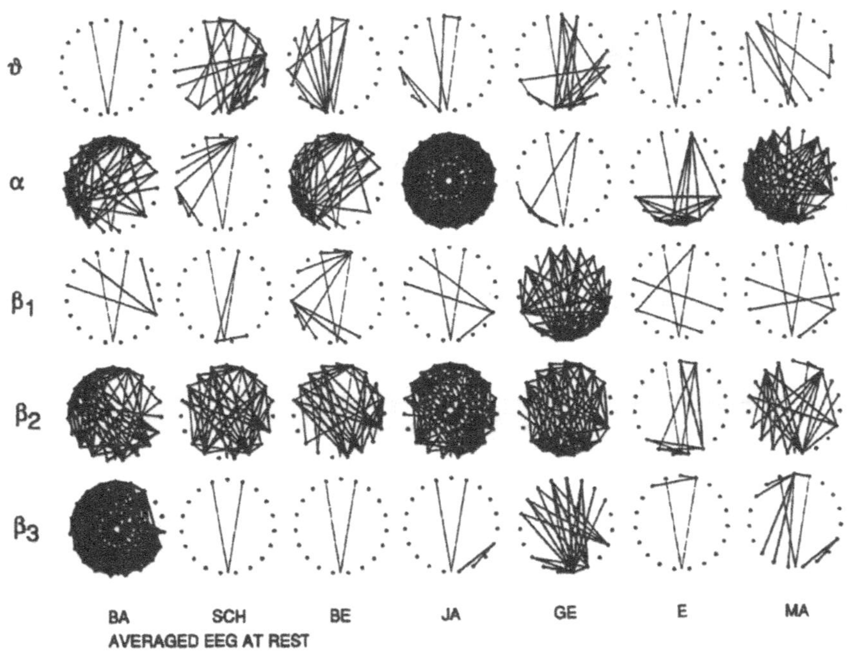

Fig.2. A 29 year-old male. Changes of coherence while different mental tasks are performed with respect to the averaged EEG at rest. From *left to right*: listening to Bach (*BA*), Schönberg (*SCH*), Beethoven (*BE*), Jazz (*JA*), a German (*GE*) and an English text (*E*) and while doing mental arithmetic (*MA*). All tasks took 5 min, except the last one (1 min).
Significant changes ($p =< 0.01$, 0.02 and 0.05) with regard to the averaged EEG at rest, which was performed for 1 min at least before, after, and between the mental tasks are indicated by lines of decreasing strengths

to chance. Such an argument may easily be invalidated by estimating coherence changes for each minute of each task; this procedure yielded hardly any differences between the five coherence patterns. Thus, most likely the coherence changes as observed are truly caused by the different natures of the tasks and one may conclude that different mental processes activate the electric coupling between different brain areas in different ways, but in a frequency-dependent manner.

A further question arises as to the possible significance of these patterns. If one assumes, according to the above conclusions and anatomical considerations, that functional aspects are reflected by these patterns of coherence changes, could it be that an accentuation of intrahemispheric coherence increases may indicate a stronger involvement of the respective hemisphere in the respective task and the respective frequency band?

Although this question cannot yet clearly be positively answered, there are some hints that, particularly in the beta bands, a greater amount of coherence increases in one hemisphere would indicate stronger involvement. This was particularly well established in studies on mental interpreting (Petsche et al. 1993a) and in other verbal listening and reading tasks (Petsche et al. 1987) in which right-handers showed more left- than right-hemispheric increases of coherence. On the other hand, the two left-handers observed thus far showed the reversed situation.

Another fact worth mentioning is that several electrode sites are characterized by participating in more coherence increases than others (Fig.3). In this representation the size of each of these squares is proportional to the number of coherence increases taking their origins at the respective places (as the largest possible number of increases taking their origin in one electrode is 18, the largest

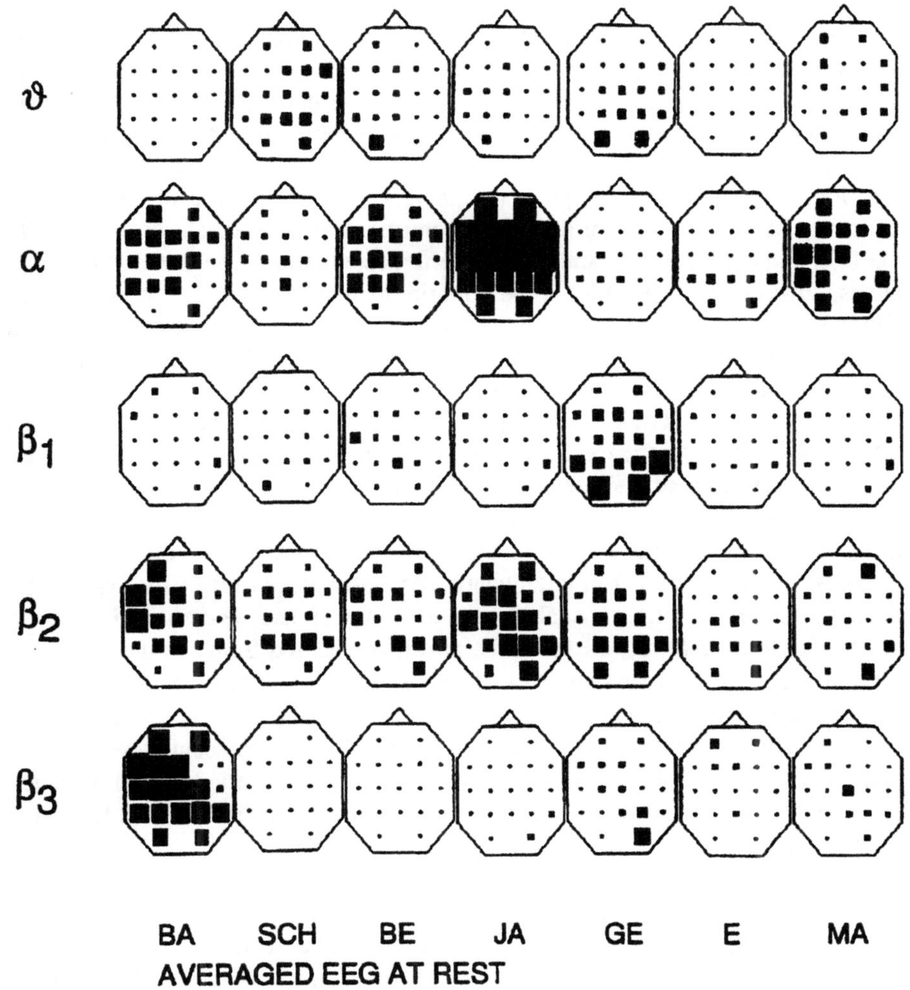

Fig.3 "Centers of gravity" of coherence increases: the size of each square is proportional to the number of coherence increases in which every electrode participates while the tasks of Fig.2 are performed. Abbreviations as in Fig. 2

square means that this electrode increased coherence with respect to all other electrodes). Such "focal areas" or "gravity centers" of coherence suggest there is a sort of nodal point for the respective function. This presumption was also supported by the studies on verbal thinking mentioned above in which the hemisphere with the larger number of intrahemispheric coherence increases in most cases coincided with the hemisphere in which the larger number of focal areas of coherence increases was found.

A further indication of the functional significance of these "focal areas" of coherence increases is that they are not randomly distributed but often form clusters, thus indicating that larger areas are involved in the performance of the respective task.

When taking all these points into account, a provisional answer to the question as to which hemisphere is most engaged in listening to music can be given: already the sample of Figs. 2 and 3 demonstrate that a clear brain lateralization of music perception cannot be assumed. Even during one experiment, when music of different styles is listened to, the hemispheric engagement may change; this is true as well for listening to the same piece at intervals of weeks. Moreover, the different frequency bands often show differences in hemispheric involvement. Most frequently the beta bands hint at the same hemispheric accentuation. The lower frequency bands, however, may also hint at the opposite hemisphere. Not much can be said about hemispheric accentuation in these bands up to now as their functional significance has not yet been clarified; not even the alpha band has turned out to be as uniform as it often is assumed to be.

The above considerations are supported by observations on musical patients after strokes, in whom the nature of a possible amusia never can be predicted from the exact knowledge of the extension and site of the brain lesion. This is in contrast to patients with aphasia, in which the probability of any sequelae concerning speech is high if either the Broca or the Wernicke region become involved.

Thus, searching for a clear lateralization of music processing by the brain could easily be the pursuit of a chimera, at least if the search is based on electrophysiological grounds. However, also if considered from a general point of view, it is hard to imagine that one half of the brain should be more engaged than the other all the time in a process as complex as listening to music. If the patterns of coherence changes may generally be conceived of as representations of some main roads of thinking strategies - and in our long-term experience with the study of different thinking processes we have come to like this idea - then the following reflections may apply.

Let us hypothesize about the differences in brain work between solving a simple arithmetic task, such as adding numbers, a verbal task, such as listening to a text and listening to music. In the case of doing mental arithmetic in this simple way, the brain probably has just to store a string of numbers and hold the sum of one addition in mind while calculating the next. It mainly operates with concepts of symbols for numbers and needs its working memory. While processing language, the brain is occupied with the transposition from thoughts into words and sentences and vice versa. To do this, the brain has access, in a special memory, to a fairly large, though limited instrumentarium of concepts, words and grammar rules in one or several languages. In both cases, a transposition is carried out between concepts and symbols for concepts, namely, numbers and some rules for numbers, on the one hand, and words and grammatical rules, on the other. In verbal thinking the neuronal networks in the brain needed for these tasks are fairly confined and most likely restricted to a relatively limited number of connections. This holds even more for solving a simple arithmetic task as in our experiments, in which the subjects are requested to add subsequent numbers only: to do this, the individual has only to use his memory of numbers and the most simple addition rules. In contrast to this, while processing music, much more diversified neuronal networks are most likely involved, and here, the question remains open as to what set of symbols the architectonic sound structures should be transposed into by the brain. At best the outcome is "pleasure," but this concept is too broad and diffuse to make any sense. Moreover, it is certainly not "pleasure" alone that motivates our strive for music. In addition, as music is experienced by everybody in a different way, it can be assumed that the two hemispheres, which are known for their heterogeneous contributions to personality, may be activated according to their inborn or learned abilities and to the present emotional situation.

These reflections have been validated experimentally. According to the above considerations the strategies the brain devises when processing music should have more degrees of freedom than those of verbal thinking, and these again more than performing a simple arithmetic task. If, furthermore, brain strategies of mental acts are reflected to a certain degree by the webs of coherence changes, then the following should apply: when repeating the same task several times during one experiment, the scattering of connections with significant coherence increases should be largest when listening to music and lowest when performing the simple arithmetic task, with a position of verbal acts in between. Moreover, the musical task should represent the lowest and the arithmetic task the largest number of increases occurring each time, again with the verbal task in between. This could be shown to be true (Petsche et al. 1993b).

It is for such reasons that many more heterogeneous coherence patterns may be expected in music perception than during other mental processing operations. In spite of this, some consistent results can be observed while processing music in the brain even if the trials are repeated at intervals of several weeks, as was demonstrated in another study.

In a study on musical expectancy (Janata and Petsche 1993) yet another auspicious application of the EEG to the study of musical thinking was demonstrated.

Expectancy plays a considerable part in enjoying music. According to Jones (1981) "expectancy schemes" reflect a listener's knowledge of how a particular piece of music should proceed in time, and thereby serve to guide the listener's attention. In the above study, explicit violations of harmonic expectancies were studied: 23 students of both sexes were presented incomplete cadences consisting of the tonic, subdominant, and dominant. The cadence was completed either with the proper (expected) resolution to the tonic, or with the ambiguous resolution to the minor or with an unexpected resolution to a chord of a harmonically most distant key.

This experiment ought to answer the question of whether changes in the EEG parameters quantifiably reflect the degree to which a musical context has been violated, and, if so, at which recording sites and in which frequency bands such differences occur. Analyses of variance showed that the two EEG parameters investigated, amplitude and coherence, and even reaction time differed between the various resolutions. The EEG parameters not only changed at recording sites located above the auditory cortices, but also at sites above right frontal and parietal regions.

Finally, what kind of brain processes do these findings represent? The brain is continuously affected by the environment via known and still unknown sensory systems. These stimuli are transformed into nervous signals and further processed in the sensory ganglia and the primary cortical areas. At this stage, perhaps also even earlier, another system becomes involved, which we prefer to call the system of differential attention, which we think tailors and pares the messages arriving at the brain so as to enable the individual to optimize the experience of his everyday encounters with the environment. After this, the events are further processed in the higher systems of brain until the appropriate outputs are constructed to react to the environment. In our opinion it is this system of differential attention to which the EEG seems to have access, and more and more it turns out that this system has a high order of complexity and participates to a high degree in those functions which we contribute to the mind.

References

Braitenberg V, Schüz A (1991) Anatomy of the cortex. Statistics and geometry. Springer, Berlin Heidelberg New York

Janata P, Petsche H (1993) Spectral analysis of the EEG as a tool for evaluating expectancy violations of musical context. Music Percept 11:117-151

Jones MR (1981) Music as a stimulus for psychological motion. Part I. Some determinants of expectancies. Psychomusicology 1:34-51

Moecks J, Gasser T, Tuan PD (1984) Variability of single visual evoked potentials evaluated by two new statistical tests. Electroencephalogr Clin Neurophysiol 57:571-580

Nunez PL (1981) Generation of human EEG by a combination of long- and short-range neocortical interactions. Brain Topogr 1:199- 215

Petsche H, Rappelsberger P, Pockberger H (1987) EEG-Veraenderungen beim Lesen. In: Weinmann HM (ed) Zugang zum Verstaendnis hoeherer Hirnleistungen durch das EEG. Zuckschwerdt, Munich, pp 59-74

Petsche H, Lindner K, Rappelsberger P, Gruber, G (1988) The EEG - an adequate method to concretize brain processes elicited by music. Music Percept 6:133-159

Petsche H, Etlinger S, Filz O (1993a) Brain electric mechanisms of bilingual speech management. Electroencephalogr Clin Neurophysiol (in press)

Petsche H, Richter P, von Stein A, Etlinger S, Filz O (1993b) EEG coherence and musical thinking 1993. Music Percept (in press)

Salinsky MC, Oken BS, Morehead L (1991) Test-retest reliability in EEG frequency analysis. Electroencephalogr Clin Neurophysiol 79:382-392

How Consistent are Changes in EEG Coherence Patterns Elicited by Music Perception

PETER RICHTER, HELLMUT PETSCHE, AND OLIVER FILZ

Introduction

One of the aims of this symposium is to emphasize that multidisciplinary research is a good approach to the study of musical cognition. In addition to a variety of psychological methods and studies of neurological or psychiatric diseases, neurophysiological approaches may further contribute to this subject, making use of the registration and evaluation of the ongoing electrical activity of the brain.

The present study is based on the premise that higher brain functions - musical thinking included - are performed by extended cortical areas in cooperation and which take much longer than a few milliseconds. Thus they are seen in contrast to elementary brain functions such as responses to simple stimuli (Leng and Shaw 1991). When mental activity is to be studied from this perspective, an investigational tool to estimate cerebral cooperation up to several minuts is needed.

Previous EEG studies of mental processes were able to show that the ongoing EEG is not just the working of chance. Instead, it proved to display information about brain processes, especially if methods are applied to compute the correlation between the electrical signals from different regions: namely, the evaluation of coherence in the ongoing EEG. Coherence is computed by means of spectral analysis (Bohdanecky et al. 1982) and shows the degree of correlation between the signals from two electrodes within a certain frequency range. It may therefore be interpreted as a measure of the functional coupling between the underlying cortical regions.

Tasks as different as listening to a text, reading, mental arithmetic, spatial imaginations, even playing chess or listening to music have been the subject of investigations using this method, primarily performed as group studies (Beaumont et al. 1978; Petsche 1991; Petsche et al. 1988; Rappelsberger et al. 1987; Shaw 1978). In all these tasks different brain regions become more or less tightly electrically coupled.

A still unanswered question in this field of research was the reproducibility of the findings for single individuals. Were they just ephemeral, according to a spontaneous functional state of the brain? Or would it be possible to achieve consistent results, if for instance one and the same person was listening to a certain piece of music at different times several weeks apart? The answer to this question would elucidate the human brain is provided with specific reproducible modes for processing musical information. This can be determined by electrophysiological means.

As a refinement of these considerations and also in order to explore the validity of the results, we decided to extend our studies: would it even be possible to point out consistent differences in cortical activity if one person was listening to various compositions - such as a polyphonic fugue or a jazz piece? As different regions of the brain seem to participate in processing certain musical elements, such as rhythm, melody or complex harmonic structures, one could assume that listening to music would activate cortical areas according to the main elements characterizing this music.

Methods

In our experiments the EEG are recorded, as in the clinic, from 19 gold electrodes (10/20 system) with averaged earlobes as reference (Rappelsberger 1989). The time constant is 0.3 s, the high frequency filter chosen at 35 Hz. Data are digitized on-line at a sampling rate of 128/s and stored on a hard disk for further processing. Artifacts, as caused by eye movements or muscle activity, are excluded by visual control.

The computation of coherence by means of spectral analysis (fast Fourier transform) is done using artifact-free periods of 2 s each. Coherence is defined by values between "0" and "1," thus representing the degree of corespondence between two signals recorded from distinct regions.

We are especially interested in changes of coherence - increases or decreases - during a task in comparison to EEG at rest. Only statistical significant coherence changes ($p <= 0.05$, confidence intervals) are considered for further evaluation. Results are shown separately for seven frequency ranges from 1.5 to 31.5 Hz, namely, delta (1.5-3.5 Hz), theta (4.0-6.5 Hz), alpha 1 (7.0-9.0 Hz), alpha 2 (9.5-12.5 Hz), beta 1 (13.0-18.0 Hz), beta 2 (18.5-24.0 Hz) and beta 3 (24.5-31.5 Hz). This is done for all possible connections between the 19 electrodes, which makes a total number of 171 possible coherence values for each frequency range.

In this paper we present the results from three test persons. All of them were clinically healthy, male, right-handed and did not take any drugs at the time of the EEG recordings. Their ages were 22, 27, and 54 years. They have played instruments since childhood and have reached a high level of performance.

During the recording sessins the subjects were sitting in a sound-proof room with closed eyes. The music was presented via headphones.

In these experiments four pieces of music were used, chosen according to the following features: all pieces last about 5 - 6 min, are written for piano solo, and are in a "moderate allegro" tempo. They vary basically in musical style:

- J. S. Bach, Prelude and Fugue XX in a-minor from "The Welltempered Clavichord, Book I"
- L.v. Beethoven, Sonata op.110, First movement

- A. Schoenberg, Piano pieces op.33a and b (twelve-tone music)
- A. Romano, "Amalgame," played by M. Petrucciani (jazz with striking rhythm)

EEG at rest were recorded for at least 2 min at the beginning and at the end of the session as well as between the four pieces. These at rest periods finally are merged. The coherence values while listening to music are compared to the values from the merged recordings at rest. Between the tasks we took a listening break of about 10 min for a short interview about the personal perception of the piece just presented.

This recording session was performed three times for each test person at intervals of about 4 weeks; the sequence of the pieces of music was random.

Results

Individual Coherence Changes Independent of the Musical Style

First we searched for common features in the coherence patterns while listening to music in general. The question was whether characteristic coherence changes for each individual could be found reproducibly through all styles of music. This actually proved to be the case.

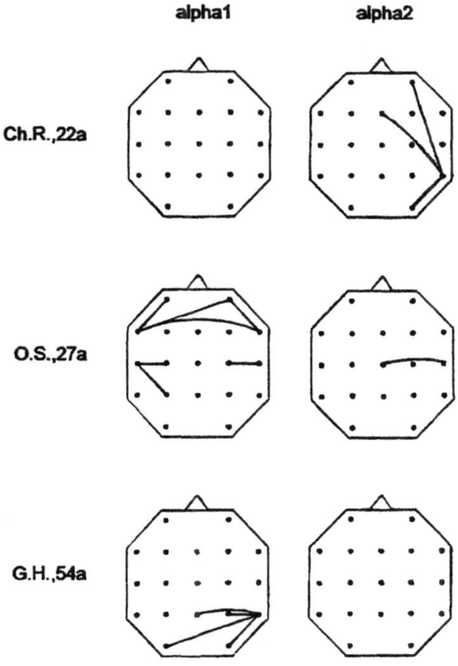

Fig. 1. Significant coherence increases ($p < 0.05$) with respect to EEG at rest, reproducible for all four pieses

Our criterion for reproducibility was that the same significant change of coherence - increase or decrease - should be found either in all three recording sessions or at least in two of them, without any tendency to the contrary in the third one.

A schematic view of the topographic distribution of all coherence connections between the 19 electrodes is shown in Fig. 1. The figure shows reproducible significant coherence increases for each person while listening to music: remarkably, coherence patterns just arising from listening to music in general, independent of the style, were found to be mainly restricted to the alpha range. Coherence changes indicated in Fig. 1 fulfill the criterion for reproducibility for all four music pieces presented in the experiment.

It turns out that regardless of whether Bach, Beethoven, Schoenberg, or jazz was presented, everyone produced his own pattern of coherence increases with regard to EEG at rest. In spite of all distinctions, the right temporal lobe always seemed to play an important role. It might be assumed that all involved areas - distinct for each individual - are significantly engaged during music listening, indeprendent of the style.

Intraindividual Differences in Coherence Pattern Depending on Musical Style

The second question whether different cortical processing during the perception of the four compositions could be distinguished with our method. For this purpose,

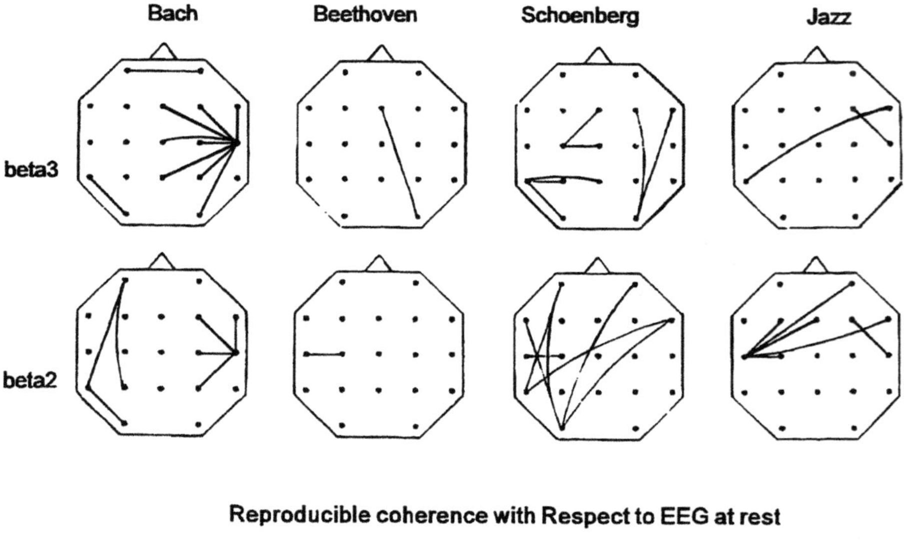

Fig. 2. A 27-year-old subject listening to all four compostions for 5 min.

again, reproducible results are demonstrated separately from the EEG recordings while listening to Bach, Beethoven, Schoenberg, and jazz. As an example, the results from a 27-year-old test person are demonstrated (Fig.2). As mentioned above, each test person produces a rather nonspecific coherence pattern in the alpha range while listening to the four pieces. The differences in beta 2 and beta 3 are more striking and are therefore presented.

From the figure an impression of the great variety of possible coherence patterns as responses of the brain on various music pieces can be formed. It is worth noticing that in every case distant areas seem to be working in cooperation and that musical perception is not limited to a single lobe. One should bear in mind that these patterns of functional cortical connections are reproduced at intervals of several weeks and therefore can be seen as rather specific for certain mental acts.

Interindividual Differences in Musical Processing

These findings lead to the next question: what ist different and what do have the three persons in common when listening to the same piece of music?

How does, for instance, Bach act upon different brains? In Fig. 3 reproducible coherence increases are shown from all three test persons while listening to Bach, again in beta 2 and 3. As expected, they vary interindividually, although some common features can be recognized.

Primarily, we find that both temporal lobes in each person are involved: They all exhibit increased functional connections between the right temporal lobe and

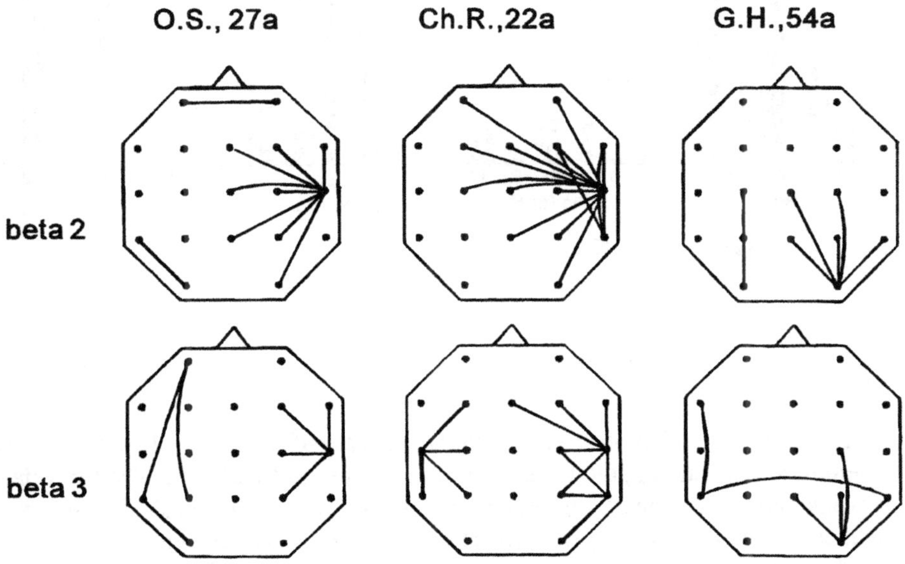

Fig.3. Significant coherence increases ($p < 0.05$), reproducible for each person while listening to Bach

right occipital areas in beta 2; additionally, temporofrontal coherence increases in the beta frequency ranges either in the left or right hemisphere. These findings demonstrate that when our test persons were listening to a fugue by Bach, the temporal lobes of both hemispheres wee consistently engaged.

Differences in the Consistency of Coherence Findings Depending on the Musical Style

The final point in our investigations concerning the reproducibility of coherence patterns was wich musical style induces the highest degree of reproducibility? This is illustrated for one case in a diagram (Fig. 4). It indicates the percentages of reproducible changes in relation to all significant changes found, separately for the five frequency ranges.

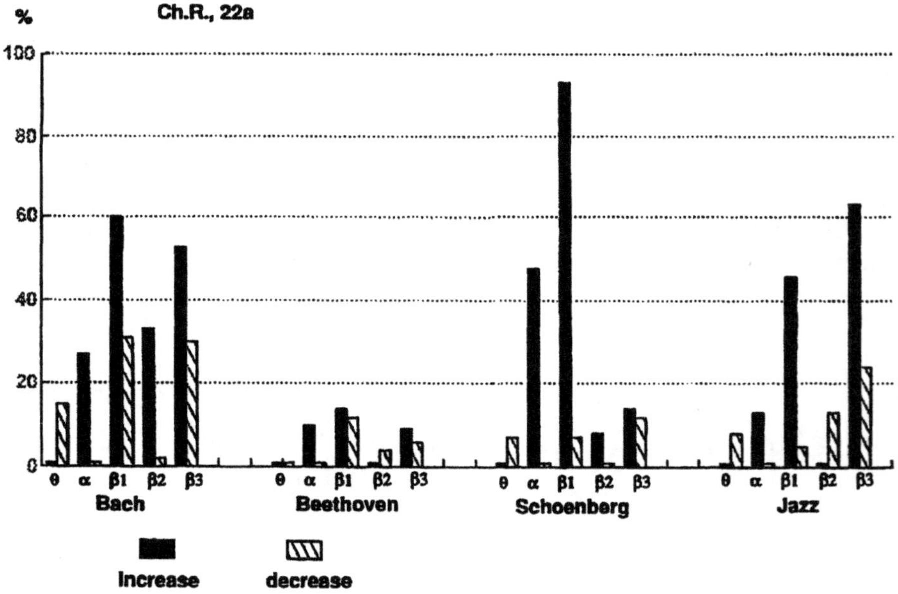

Fig. 4. Percentage of reproducible significant coherence changes ($p < 0.05$) in a 22-year-old subject

This test person demonstrates the highest degree of reproducibility when he is listening to Schoenberg, the lowest one for listening to Beethoven. The two other persons show the highest reproducibility while listening to Bach. In all cases increases of coherence are generally better reproduced than decreases.

How might these considerable differences be explained? One possible reason might be the listener' s attitude towards the music presented. When asked about their intellectual interest and emotional empathy to each piece, the listener' s intellectual interest in a certain music piece seemed to be the determining factor for good reproducibility of coherence patterns. Thus, a higher degree of concentration could be one reason for more specific mental processing and therefore give rise to coherence patterns which are reproduced better.

A Comparison Between Composing Music and Listening to Music

As a supplement to the previous observations we had the chance to evaluate EEG recordings from one of the three test persons while he was composing a piece of music, so it is possible to compare a creative task to the former cognitive tasks of listening to music.

The experimental setup was the same as before: the subject was sitting with closed eyes but composed a short tonal piece, a three-part fugue (Fig. 5). He was able to do this by hearing the music mentally without visualization of the score. It took him five minutes on the whole. Because of his excellent musical memory he was able to note it down immediately after the EEG recording.

The analysis in Fig. 6 comprises the EEG of the total time of composing. Looking at coherence changes with respect to the merged EEG at rest, a dense net of increasing connections is found, extending to all regions. The results are demonstrated in three ranges, delta, alpha 1, and beta 2, in order to show that slow

Fig. 5. Three-part-fugue composed by one subject during the EEG recording

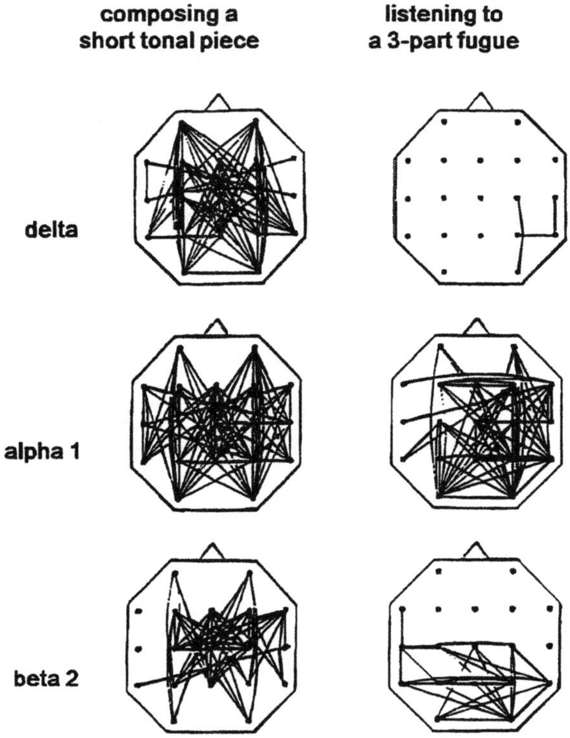

Fig. 6. Significant coherence increases ($p < 0.05$) while composing music (**a**) and while listening to Bach (**b**)

as well as fast waves are involved in this act of creating music. The striking amount of interhemispheric connections during musical composing in this case should be mentioned. Compared with coherence changes arising from listening to music, a specific feature of the creative task is the intense engagement of low frequencies, namely, in the delta range. This phenomenon seems to be specific for this person, because similar results were achieved four times when we repeated EEG recordings while he was composing.

Summing up, striking diffeences between composing and listening to music is the extension of coherence increases to larger cortical areas as well as the reinforcement of interhemispheric connections while composing a piece. This leads to the assumption that - in this particular case - the creative task of composing involves even more and different parts of the brain than perceiving music.

Discussion

From these findings it becomes evident that considerable intra- and interindividual differences exist in cortical processing during the perception of music, as displayed in coherence pattern. Thus, one cannot speak of special "musical processing areas." Even for individual persons consistent results, independent of the presented piece, are only found in the alpha range, which means that cortical processes reflected in all other examined frequency ranges (especially beta bands) are rather specific for distinct musical styles. Yet one common feature in coherence changes elicited by music perception is figured out for all test persons: the frequent engagement of the right temporal lobe, either in alpha 1 or 2.

Taken as a whole, the cortical involvement in the perception of music, as displayed by coherence patterns, fits quite well to findings from other studies. As discussed by Gordon (1983), distinct musical elements appear to be represented in different regions of both hemispheres, yet not at all stable, but rather varying from person to person. According this conclusion, the perception and further processing of a certain piece of music would involve both hemispheres and therefore naturally need cooperation between "left" and "right." We found corresponding results in our study, with emphasis on the importance of the right temporal lobe for all four musical styles we examined. Finally, the examination of composing music, showing the contribution of all cerebral lobes to this creative mental act, supports our earlier statement that not a single gyrus or even lobe can be responsible for musical thinking. This aspect of personality - as it seems - is set up by a variety of abilities in cooperation, coordinated by an overriding functional state of the brain.

References

Beaumont JG, Mayes AR, Rugg MD (1978) Asymmetry in EEG alpha coherence and power: effects of task and sex. Electroencephalogr Clin Neurophysiol 45:393-401

Bohdanecky Z, Lansky P, Radil T (1982) An integral measure of the coherence function between pairs of EEG recordings. Electroencephalogr Clin Neurophysiol 54:587-590

Gordon HW (1983) Music and the right hemisphere. In: Young AW (ed) Functions of the right hemisphere. Academic, London, pp 65-86

Leng X, Shaw GL (1991) Toward a neural theory of higher brain function using music as a window. Concepts Neurosci 2:229-258

Petsche H (1991) The information content of the human EEG. In: Haken H, Koepchen HP (eds) Rhythms in physiological systems. Springer, Berlin, Heidelberg, New York, pp 257-272

Petsche H, Lindner K, Rappelsberger P, Gruber G (1988) The EEG: an adequate method to concretize brain processes elicited by music. Music Percept 6(2):133-160

Rappelsberger P (1989) The reference problem and mapping of coherence: a simulation study. Brain Topogr 2:63-72

Rappelsberger P, Krieglsteiner S, Mayerweg M, Petsche H, Pockberger H (1987) Probability mapping of electroencephalogram changes. Application to spatial imagination studies. J Clin Monit 3:320-322

Shaw JC, O'Connor K, Ongley C (1978) EEG coherence as a measure of cerebral functional organisation. In: Brazier MAG, Petsche H (eds) Architectonics of the cerebral cortex. Raven, New York

Motor Cortical Fields and Rhythmic Coordination of The Hands in Music Performance

LÜDER DEECKE

Introduction

This chapter will not be dealing exclusively with music but with general motor and skill performance, which includes playing music. We are interested not so much in the sensory side (listening to music) but more in the motor side (playing music), e.g., when playing rhythms. When an individual makes a voluntary movement, such as moving the finger, which is also important in music playing, when, for example, pressing down a key on a piano or a string on the fingerboard of a violin or cello, then the brain prepares for this movement. By recording the EEG of such a subject and focusing on the DCwaves, we can record from the brain the signals in conjunction with such preparation for movement. Many signals are contained in the EEG: Evoked potentials following an event, coherence patterns that accompany music processing (Petsche et al., this volume) or DC-EEG in music processing (Beisteiner, this volume). Our group has been interested in slow potential shifts preceding voluntary movement for 30 years now (Kornhuber and Deecke 1964, 1965). The negative shift occurring under these circumstances is called the *Bereitschaftspotential* (BP) or readiness potential, which starts as early as 1 to 2 s prior to the onset of movement. Negativity implies that the cortical area under study is active, positivity means that it is inactive.

Potentials Preceding Voluntary Movement

Figure 1A shows the potentials accompanying voluntary movement. At point zero (vertical bar in Fig. 1A), the subject - a patient with Parkinson's disease in this case - starts performing a voluntary left finger movement and she is already preparing for it 1 s or more before the onset of movement by the BP. What also can be seen is that over the lateral motor cortex (precentral gyrus hand region, area 4, "primary motor cortex") no particular BP is seen. However, over the vertex, which is in the midline, a large BP can be seen. This finding suggests that the BP is not generated from the convexity of the two hemispheres - as there is no BP - but rather from structures that are in the midline of the brain between the hemispheres. These

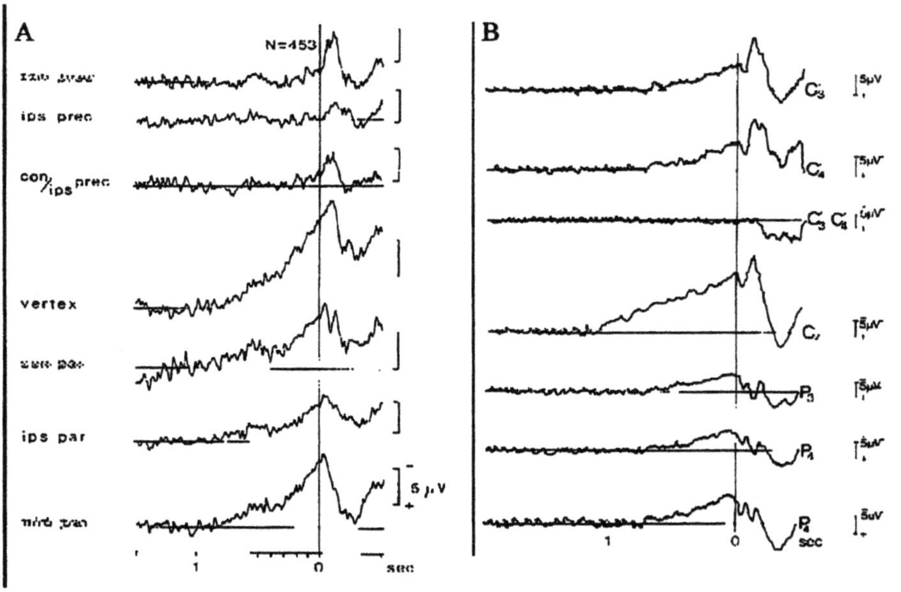

Fig. 1. **a** Cerebral potentials preceding voluntary movement in a patient with Parkinson's disease. Averages of 453 left index finger flexions recorded in contralateral and ipsilateral precentral and parietal leads (*con prec, ips prec, con par, ips par and mid-parietal, mid-par)* and the *vertex*, referred to linked ear lobes. Note maximal Bereitschaftspotential (BP) at the vertex, marked negativity also over parietal areas, but virtually no BP in precentral leads overlying the hand representation of the primary motor cortices (MI). Thus, the vertex maximum cannot be the result of potential pick-up from both MI cortices but must have its own potential generator, which is the mesial frontocentral cortex including the supplementary motor area, SMA. From Deecke and Kornhuber (1978).

Fig. 1. **b** Cerebral potentials preceding bilateral index finger flexions. Note earliest activity (about 1.1 s prior to movement onset in the EMG) in C_z - collecting activity from the SMAs. Only later (about 0.7 s prior to movement onset) does the BP occur over the primary motor cortices (MI; recordings C_3' and C_4' (*upper traces*). Modified from Kristeva and Deecke (1980)

mesial cortical areas contain a region that is known also to participate in the preparation of a voluntary movement, the supplementary motor area (SMA). It is generally believed that Penfield coined this term; however, when going back to the old German-language literature, and in particular to Oskar and Cecilie Vogt (1919), we realize that they already distinguished in the chimpanzee the term *Motorische Supplementärarea*, or motor supplementary area. Otfried Foerster (1934) continued the tradition and described a SMA in man by stimulation mapping. Penfield (1950) - as Foerster's disciple - introduced the term into the English-speaking world. The present state is that we have at least two areas in the cerebral cortex that deal with motor, the precentral gyrus' area 4 (primary motor cortex, MI) and the SMA (see

Deecke 1987). In our paper on Parkinson patients performing voluntary finger movement, where Fig. 1A was taken from, we proposed SMA participation in preparing for a voluntary movement (Deecke and Kornhuber 1978). Confirmation of this notion came from the group of Lassen et al. (1978), who published data on regional cerebral blood flow (rCBF) with the xenon intracarotid injection method while a subject was making movements with his right hand. They found that not only was the hand area of the primary motor cortex active, but also the mesial SMA

This was the situation with unilateral movements; how is it with bilateral movements? This is important for musicians because on the keyboard and with most instruments we are always dealing with bilateral - bimanual motor activities. The organ or percussion even require bipedal in addition to bimanual motor skills. We all know that playing music demands highly sophisticated motor coordination. In Fig. 1B, from Kristeva and Deecke (1980), bilateral movements of the index fingers are performed. When considering the recording C_Z which is at the vertex, the middle of the skull, we see the earliest beginning of the BP, already starting 1 s before movement onset. At the MI of both sides C_3 (left) and C_4 (right), the BP starts later, i.e., only 0.5 s prior to movement onset. Thus, it became obvious that it is probably not the MI that starts the movement, but rather the SMA at the midline C_Z that has the starting function. Thus, SMA is "upstream" in the chain of cortical activations preceding voluntary movement (see, for example Deecke et al. 1990). Although this is still a matter of debate, we will see (see the section *Potentials with Lesions of the SMA*), piece by piece more accumulating evidence that this order of events - first SMA, then MI - obviously is the rule when we prepare for voluntary action. In the end it might become clear that it is the SMA that channels motivation and intention into execution of movement (see Deecke and Lang 1990).

Potentials related to motor learning

Another very important question that is of interest for playing music is how do we learn new motor skills. The experiments on learning were done with Wilfried and Michael Lang (Lang et al.1986). They designed an experiment which also involves the SMA and which also has some bearing on playing music, e.g., reading music (Fig. 2A). The subject looks at a televisionscreen and has a light stylus in his right hand. When he moves it on to a photodetector plate, he starts a circle on the TV screen to move for 1.5 s in a first direction, then for 1.5 s in another direction, and finally for 1.5 s in a third direction, thereafter resetting to the original position. The time constraint was always constant, i.e., predictable; 1,5 s per trajectory. However, the space constraints were absolutely unpredictable, since the directions were created by a random generator. As soon as the light circle moved on the screen (Fig.2), the subject was to track the moving circle on the screen by the light stylus on the photodetector plate. In order to check whether the subject´s tracking was accurate, we coupled the position signal of the stylus back to the TVscreen as a light spot (Fig.2A). Thus, accurate tracking means that the light spot is in the center of the

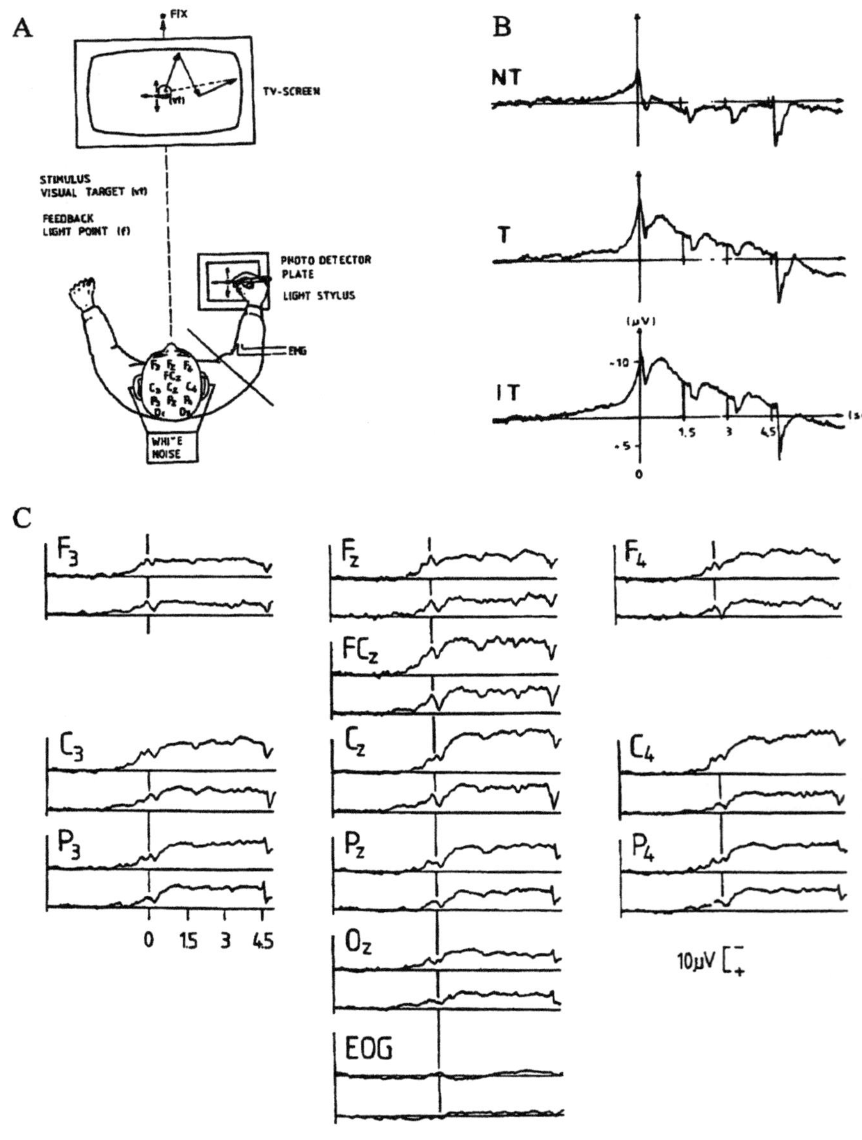

Fig. 2. a Experimental setup for the motor learning paradigm. Subjects tracked the visual target (*circle*) with the light stylus in their right hand. The position signal of the light stylus (horizontal and vertical coordinates of tracking movements) is coupled back to the TV screen as light spot. In the learning paradigm (inverted tracking, IT) the position signal of the stylus was multiplied by -1 when coupled back to the TV screen. Thus, when the circle on the screen moved in one (randomized) direction, the subject had to track in the opposite direction in order to keep the light spot within the center of the light circle. (From Lang et al. 1986).
Fig. 2. b BP and tracking-related potentials for no-tracking (*NT, top*), tracking (*T, middle*), and inverted tracking (*IT, bottom*) tasks. AC recordings (time constant = 5s). (From Lang et al. 1983).

Fig. 2. c Averages of tracking-related potentials (*T, bottom graphs* per electrode position) and learning-related potentials (*IT, top graphs* per position) across all 15 subjects. DC recordings (=). The *vertical line* indicates the volitionally initiated onset of the stimulus program (t=0). Stimulus directions changed at t=1.5 s and t=3 s, end of stimulus program at t=4.5 s. Negative up. Note that the volitional initiation of stimulus onset is preceded by a BP, task performance is accompanied by a plateau negativity, called specifically tracking potential or generally negativity of performance (N-P). (From Lang et al. 1988a.)

light circle. And when it deviates from the center, such tracking error can easily be measured electronically and can serve as a measurement of subject's tracking ability. This was the situation with normal tracking (T) serving as our control task. In the same experiment, we introduced a learning task, which was designed as inverted tracking (IT). In this task, the position signal of the tracking light stylus was multiplied by minus 1 when coupled back to the TV screen, i.e., the subject had to track the trajectory of the light circle in the inverse (mirror) direction in order to keep the light spot within the center of the circle. We told the subjects that we would check the accuracy of their tracking and that good performance would have an impact on their hourly rate of pay. So subjects were really motivated and tried hard to learn and perform the inverted tracking, which is not routine in everyday life, but has to be learned.

Figure 2B gives the brain potentials at electrode position FC_Z, overlying the SMA. These are data derived from AC amplification (time constant = 5 s); we see a drop in negativity over the time period. At that time we thought that the negativity (= activity) of the brain returns to zero during the subject's tracking of the trajectories. When our new equipment was ready, we were able to record with real DC amplification (=). This is shown in Fig. 2C: We see that there is sustained negativity during the entire tracking time and no drop at all, proving that the drop in Fig. 2B was an artifact due to the time constant. In reality the cortex stays active during the tracking, which - if we make again inferences to our musicians - would also be the case when reading music.

In Fig. 2C we see two graphs per electrode position. The upper graph is always more negative than the lower graph, particularly over the SMA. The upper graph is the inverted tracking IT, the learning task, and the lower graph is the control task, the normal tracking T. If we take the difference between IT and T, we have a measurement of the extranegativity in the learning task over the control task. This extranegativity may be related to learning itself.

In Fig. 3A, we correlated the extranegativity for learning (i.e., negativity in the learning task minus the one in the control task) with the deltaE, which is the tracking error in percent (measured as the deviation of the light spot from the center of the light circle, see Fig. 2A). We found significant correlations between the reduction in error (= success in learning) and the increase in negativity of our 14 subjects, in the sense that there were good learners and bad learners. What is interesting, however, is that the good learners generated more negativity and the bad learners generated less. These correlations were significant exclusively in the three *frontal* electrode positions F_3 (frontal convexity of the left hemisphere), F_4 (frontal convexity of the

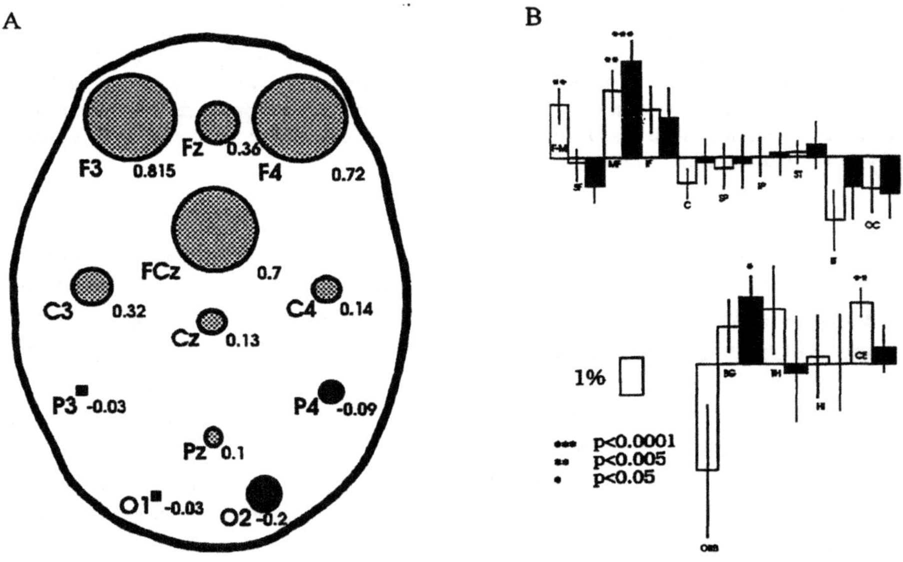

Fig. 3. a This map is derived from correlations done on the data of the motor learning experiment of Fig. 2. The extra-negativity for learning (IT minus T) was correlated with the success in learning (minimization of error E during the experiment, measured as the deviation of the light spot from the center of the light circle, see Fig. 2A). Significant correlations (coefficients of correlation r between 0.82 at F_3 and 0.7 at FC_z) are seen over frontal cortex only (F_3, F_4, FC_z). In the map not *r* but r^2 (coefficient of determination, *d)* is plotted as the area of circles. Note that it is the frontal brain that is involved in motor learning (which would also apply for learning a musical instrument).(From Lang et al. 1983).

Fig. 3. b This plot is derived from the regional cerebral blood flow, rCBF experiment (emission tomography, SPECT using the tracer ^{99m}Tc-HMPAO) carried out with the same subjects as in Fig. 2 and 3a. The difference between inverted tracking (IT, learning task) and normal tracking (T, control task) is plotted per region of interest (ROI). ROIs of the midline are *white*, those of the left side are *light gray* and those of the right are *dark gray*: Note that also with this method it is the SMA (*F-M*, frontal-medial region, extreme *left white column* in *top row*) that shows extra activity for the learning task (IT, $p<0.005$). Even larger learning-related extra activity is seen - corresponding nicely to Fig. 3a - in the frontal convexities (middle frontal [*MF*] gyrus of the left hemisphere (*light gray MF* column, $p<0.005$) and particularly the right hemisphere (*dark gray MF* column, $p<0.0001$). Note learning-related extra activity also in the basal ganglia (*BG*, *bottom row*, the right BG, dark gray, being larger - $p<0.05$ than the *left BG, light gray*) and cerebellum (*CE*, *extreme right columns*, *bottom row*). Due to the crossing fiber connections, we now find an opposite side difference, *left CE* (light gray, $p<0.005$) being larger than *right CE* (*dark gray*). (Modified from Lang et al. 1988a) right hemisphere), and FC_z (mesial frontocentral midline, SMA), as is seen in Fig. 3B which illustrates these correlations graphically. The coefficient of correlation r is not shown, but rather the coefficient of determination d (= r^2) is plotted as the area of circles. Significant correlation between learning success and extranegativity is found over frontal areas only (F_3, F_4, and FC_z).

The DC potential method is not the only way to record and topographically map cortical activity, emission tomographic techniques can also be used to do this although with a much lower time resolution. At our clinic, single photon emission computerized tomography (SPECT) was available and was carried out using the same population of subjects that participated in the DC potential study (Lang et al. 1988a). HMPAO was used as a tracer. It has the effect - similar to that of deoxyglucose in positron emission tomography (PET) - that it is trapped in brain tissue. This means, when one injects the tracer substance without the subject knowing at the culmination point of an experiment when he or she is really in a "set," one gets the pattern of cerebral blood flow - and this is to almost 100 % equivalent with cerebral activation - prevailing in the particular experiment under study. The tracer is trapped in the brain and does not change for at least 2 h (steady state). Theoretically, one could even put the subject under general anaesthesia after the injection and still would not get a different pattern under the scintillation counter. Because of this trapping effect one is able to "freeze in," so to say the cortical activity that was prevailing during the experiment under investigation.

In Fig. 3B, the difference between inverted tracking (learning task, IT) and normal tracking (control task, T) is plotted per region of interest (ROI), and we see that also with this method it is the SMA (F-M, frontal-medial region) that shows extra activity for the learning task (IT), and also the frontal convexity (middle frontal, MF gyrus of the left and particularly the right hemisphere). On SPECT other brain areas besides the cortex can also be investigated, and we also see learning-related extra activity in the basal ganglia (BG, R>L) and cerebellum (CE, L>R). The side difference is interesting: The right MF region shows more learning-related extra activity than the left MF region. So the SPECT data nicely confirm our DC-EEG results. In terms of statistics, the SMA extra activity is significant at $p<0.05$, the one of the left MF region is at $p<0.005$, and that of the right MF region even $p<0.0001$. The right hemisphere preponderance fits very well, because a tracking task is a spatial task, a constructional task, so to speak, which involves more the right hemisphere. This is the expert for our spatial-constructive abilities, and, by the way, also for music. Regarding hemispheric differences we usually underestimate the abilities of the right "nondominant" hemisphere. I think we have to conceive that both hemispheres are always involved in a task. However, if we divide our accoustic world into verbal sounds and nonverbal sounds, the left hemisphere is more involved in the verbal sounds (speech), and the right is more involved in the nonverbal sounds (music). As is further seen in Fig. 3B, also the basal ganglia show this right-hemispheric preponderance: The right basal ganglia generate more learning-related activity than the left. And at the level of the cerebellum (CE) it is of course crossed, because the right cerebral hemisphere corresponds with the left cerebellum and the left hemisphere with the right one.

Potentials with Lesions of the SMA

In two large studies (Deecke et al. 1987, Asenbaum et al. 1990) we investigated the cerebral potentials preceding voluntary movement in patients having lesions of one SMA. A doctoral student (W.J. Heller) went through 20 000 cranial computer tomograms, picking out lesions that were good for our question, i.e. it had to be a solitary lesion in exactly the region where we wanted it, namely in one SMA, and the rest of the brain should be intact. The lesions included vascular insults of branches of the anterior cerebral artery, but there were also cases after operation of meningiomas (which are benign tumors) of the falx, which is the sickle-shaped septum between the two cerebral hemispheres. All patients were in a chronic state several months after the lesion. Thus, we cannot say how they behaved in acute situations, but in the chronic we can: As is seen in Fig. 4A, there were no differences between the affected crossed hemisphere-hand system and the healthy one (see legend to Fig. 4A), albeit disappointing for us investigators. However, for the patients this is a good message: In the chronic state simple finger movements such as simple index finger flexions are not affected by the loss of one SMA. Since we have two SMAs, the other, healthy SMA can obviously compensate for the loss of one. However, we had the suspicion that at this time we did not ask the real "SMA question." The SMA may be needed for performing more sophisticated movements and not of simple movements. We did not ask whether the patients played an instrument, and if so how they were doing after the SMA lesion. Such a study would be of utmost interest!

In order to investigate the "SMA question," Wilfried Lang in our group in Vienna designed a splint of the hand and arm with goniometers for monitoring joint position (Lang et al. 1988b). We first investigated more complicated movements in healthy subjects, including one which may be important for music playing. The question we asked was, is the SMA more involved when movements are of temporal complexity or is it more involved in the planning and performance of spatial complexity. Under comparison were bimanual index finger movements in the same direction versus those in opposite directions - this being a spatial difference - and simultaneous movements versus sequential ones - being a temporal difference. The result showed that healthy people can easily perform these tasks with flawless recordings of the goniometers be they simultaneous or sequential. We will see later that this was different in patients.

Concerning the brain potentials preceding the task (BP) and accompanying the task (negativity of performance, N-P) it was obvious particularly over the frontocentral midline that the spatial constraints (movements in same or opposite direction) did not cause much difference. However, in the time domain there was significantly greater negativity from the SMA with the sequential as compared to the simultaneous task. In the sequential task, there was a plateau of negativity over SMA, while in the simultaneous task, negativity of performance returned to base line already during the performance, although we were recording with real DC (=). When sequential movements were compared to "uncoordinated movements," the difference between sustained negativity in the sequential task and the return to

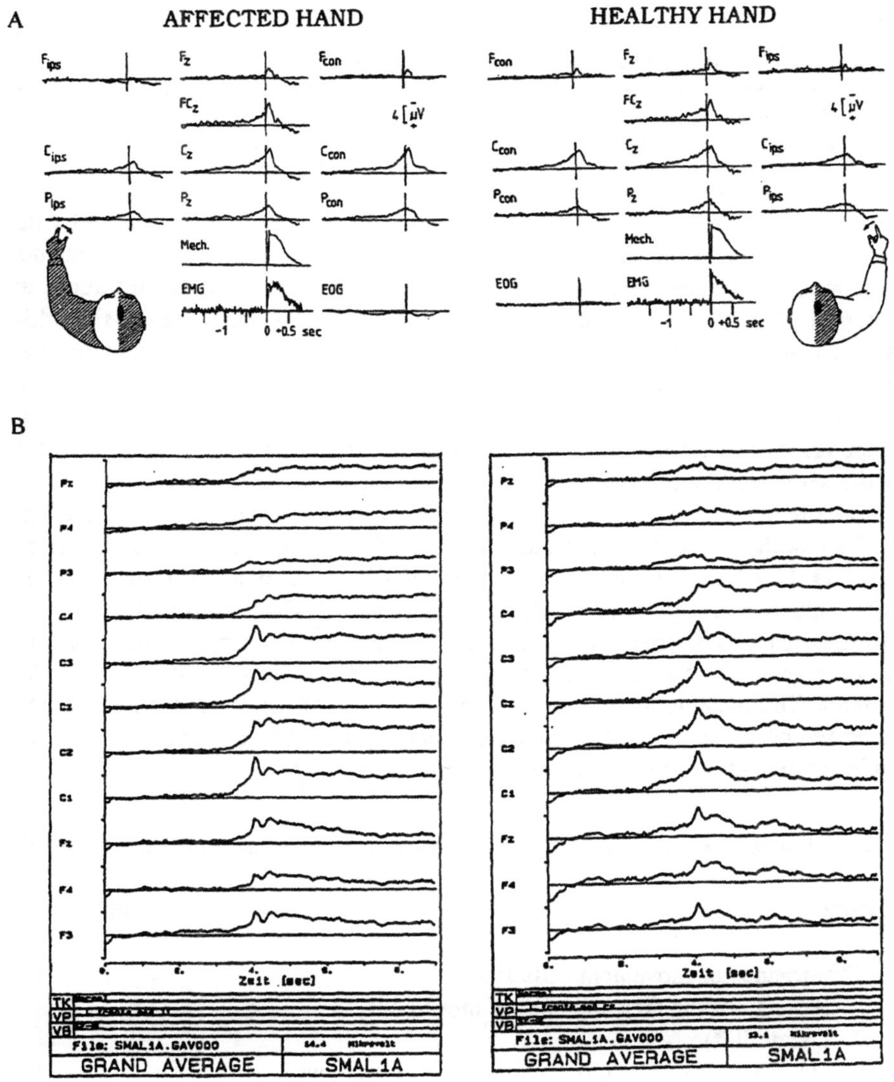

Fig. 4. a BP in SMA lesion I: Patients with unilateral SMA lesions performed unilateral simple index finger flexions with their healthy hand contralateral to the intact hemisphere and their affected hand contralateral to the hemisphere carrying the lesion. The data (grand averages across 12 patients) were organized in terms "affected hand" (*left*) and "healthy hand", (*right*) Affected hand means the affected crossed hemisphere hand system (CHHS, *hatched* in the schematic illustrations); "healthy hand" means the healthy CHHS (*white* in the illustrations). Lateral electrodes relabeled: Frontal leads $F_{3,4}$ became F_{ips} or F_{con} depending on whether they were ipsilateral or contralateral to the movement, same with central $C_{3,4}$ and parietal $P_{3,4}$ leads. Note that when comparing C_{con} of the affected hand (left) with C_{con} of the healthy hand (right) there is virtually no difference in the BP, nor is there when comparing the other corresponding leads.(From Deecke et al. 1987).
Fig. 4. b BP in SMA lesion II: At the *left*, grand averages of DC potentials of the one group of seven patients are shown having left frontomedial (SMA) lesions; at the *right* are the DC potentials in right frontomedial (SMA) lesions. Both perform the sophisticated bimanual motor coordination

task (simultaneous versus sequential movements) descibed in the text. Both groups showed larger (especially sustained) negativity for the sequential task (shown) as compared to the simultaneous task (not shown) similar to the normal subjects. Note when comparing the two hemispheres, at the left with the left-sided SMA lesions, C_3 is larger than C_4, and at the right with right-sided SMA lesions, C_4 is larger than C_3, i.e., the patients have the larger potential over the hemisphere that bears the lesion, for explanation see text. (From Asenbaum et al. 1990)

zero with uncoordinated movements was particularly dramatic, i.e. the latter do not need the help or the surveillance of the SMA, while the sequential movements do (temporal coordination). They depend more on the SMA than the difference in direction of the movements do (spatial coordination), too. Thus, we consider this as evidence for a possible timing role of the SMA which is more important than the role in spatial coordination of the two hands. This has some bearing on music performance, one could guess that the SMA is particularly challenged by different tempo patterns, more than by spatial patterns that could more or less be carried out by the MI motor areas alone.

The other study of SMA lesions was done in Vienna. Again a doctoral student (B. Oldenkott) selected out of a large number of CT scans (30 000) seven patients who had their lesion in the left hemisphere and seven patients who had it in the right. This second population of patients with chronic SMA lesions was confronted with the difficult movements as described above with healthy subjects (simultaneous versus sequential). We found that they were not able to do it so flawlessly as the normals. The patients made typical mistakes in performance of this task, which tested temporal versus spatial bimanual coordination. The mistakes the patients made can be put into three categories, as seen in Table 1 .

Table 1. Disturbance of movement performance with simultaneous vs. sequential finger movements:

Category	Number
I. Disturbance of movement initiation	
Slowing of the first segment of movement	7
Delay (left)	9
Omission of one or more segments of movement	
Unilateral (left)	8
Unilateral	5
Break off	4
II. Disturbance of movement rhythmicity (*motor dysrhythmia*)	
Transition into simultaneous action	5
Disturbed right/left alternation	4
III. Disturbance of movement performance	
Contralateral bradykinesia	11

Of particular importance is category II, *motor dysrhythmia*. This is the term we propose for the clinical syndrome governed by the lack of temporal coordination of the two hands. This is particularly important in musicians. So far we have not encountered an SMA lesion in a musician. However, the prediction would be that he would have difficulties in temporal bimanual coordination, which would critically impair playing with both hands.

For the clinicians among the readers, category III may be interesting: It is well known that bradykinesia is one of the key symptoms of Parkinson's disease. However, these patients did not have Parkinson's disease. Recently, it has increasingly been acknowledged that Parkinson's disease and frontal brain lesions have striking similarities, and that we can also find bradykinesia in frontal lesions. In our opinion this is a deficit in voluntary movement initiation in both - Parkinsonism and (mesial) frontal brain lesion (see Lang et al. 1990a; Goldenberg et al. 1990).

Figure 4B gives the results in terms of brain potentials in the two groups of SMA lesions (sequential task only). At the left are the DC potentials in left frontomedial (SMA) lesions, at the right the potentials in right frontomedial (SMA) lesions. The results of the simultaneous task are not shown, but as in the normals both groups had larger (especially sustained) negativity for the sequential task as compared to the simultaneous task. When we compare the two hemispheres in Fig. 4B for the sequential task, we find that they have the larger potential over the hemisphere that bears the lesion: In Fig. 4B right side with the right frontomedial (SMA) lesions, C_4 (which is on the right hemisphere) has more negativity than C_3, and in Fig. 4B left side with the left hemisphere SMA lesions C_3 (which is on the left hemisphere) shows more negativity than C_4. This is paradoxical: The negative DC potential is larger (more activity) on the side of the lesion rather than on the side of the healthy hemisphere!

The explanation for this seemingly paradoxical result is the following: The active site is the remaining, healthy and active SMA. This is located on the mesial surface of the hemisphere facing the falx. Thus, it creates a dipole, whose negative pole is pointing towards the opposite hemisphere, i.e., the one that bears the lesion. We observed earlier when comparing finger movements with toe movements (Boschert et al. 1983) how dipoles on the mesial surface of a hemisphere behave: With finger movements, a contralateral preponderance of negativity (CPN) is found, i.e., if we make right finger movement, the BP is bilateral but the left is larger than the right in the second half of the premovement period (BP2). However, with toe movements, it is the other way around, i.e., when we make a right toe movement, the right hemisphere has the larger BP2 as compared to the left. How can this be? If we imagine Penfield's (1950) motor homunculus, the hand is somatotopically represented on the convexity of the precentral gyrus, while the foot is represented on the mesial surface of the gyrus between the hemispheres, facing the falx - just posteriorly of the SMA, by the way. If the hand makes movements, the hand area becomes active prior to it (BP), creating a dipole - surface negative, depth positive - pointing towards the hemisphere contralateral to the movement, thus generating a CPN. However, when the foot makes a movement, the BP dipole is on the mesial surface of the hemisphere pointing towards the other hemisphere, namely, to the one that is ipsilateral to the movement (creating an ipsilateral preponderance of negativity, IPN). This stereo arrangement of dipoles can explain the paradoxical findings of both foot movements and also SMA lesions. More recently, Lang et al. (1991) investigated a patient with a right SMA lesion using magnetoencephalography (MEG). The patient performed voluntary flexions with his right index finger. In the early phase of the readiness magnetic field (0.8 to 0.6 s prior to the onset of movement) field lines were going out of the head at the vertex (C_Z) and were going into the head at a frontal position between F_3 and F_Z. They thus enveloped an electric dipole on the mesial surface of the left hemisphere in the left (intact) SMA. In the late phase of the readiness field (during BP2) between 0.2 and 0 s prior to movement onset, activity had shifted: Magnetic field lines then went out of the head at FC_3 and into the head at FC_Z. This set of isofield lines indicated an electric dipole in the left area 4 hand representation (MI). This again indicates that the SMA leads the MI motor cortex activation prior to human voluntary movement as has been described earlier in this paper, when we discussed Fig. 1B (see above).

Potentials Accompanying Rhythmic Coordination of the Hands in Musical Performance

We are now in a position to understand our music experiment. We carried out experiments (Lang et al. 1990b) in music students of the *Wiener Musikhochschule* as our subjects (Roland Beisteiner, this volume, was one of them). Differences in tempo (rhythms) between the two hands are an especially important feature of playing music, and respective experiments could help us understand the function of the SMA. The pianists among the readers might know how easy it is to play eight notes in one hand and also eight notes in the other, but how difficult it is to play eight notes in the one hand and triplets in the other! Part of the data were presented at the 4th Herbert von Karajan Symposium in Vienna (Deecke and Lang 1989). In Fig. 5A, the rhythmic tapping tasks - tapping was done with the index fingers - are shown in musical notation: Upper row; right hand,. simple rhythm: (RH-S). The right hand begins and the left hand joins in tapping eight notes with both hands; second row, complicated rhythm: (RH-C); the right hand begins with eight notes, and the left hand joins in with triplets; third row, left hand, simple rythm: (LH-S), the left hand starts and the right hand joins in with the simple rhythm (eight notes with both hands); bottom row, LH-C, the left hand starts with eight notes, and the right hand joins in with triplets.

Figure 5B gives the brain potentials in a typical example LH-S (right) and LH-C (left) of a subject (music student). Both tasks are preceded by a BP. Let us first consider the right side of Fig. 5B: We see at the bottom in the goniometer readings that the left hand starts and the right hand joins in with the simple rhythm (eight notes with both hands). The instant the RH joins in, the brain potentials show only little negativity - not very spectacular. However, at the left side of Fig. 5B with the complicated rhythm, when the RH joins in with the triplets, even professionals

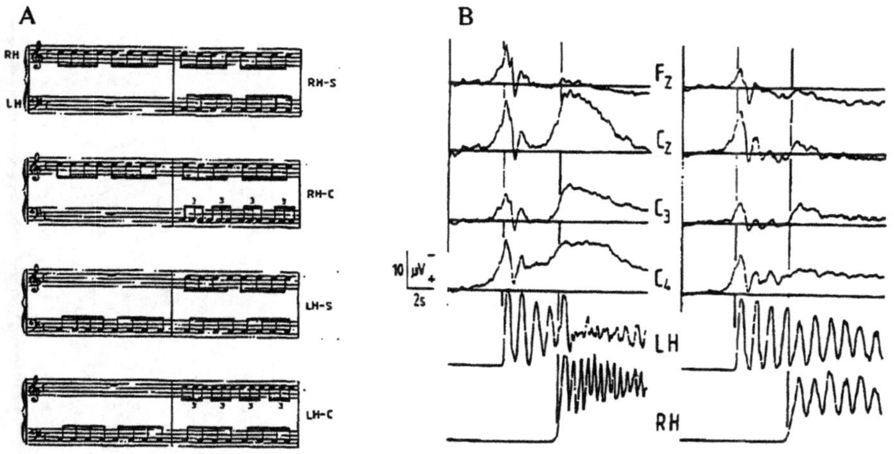

Fig. 5. a Music experiment. The music notation describes the four conditions of the rhythmic tapping task (as if hitting keys on a piano): 15 musical students tapped with their index fingers at their own volition. *RH-S* (right hand, simple rhythm): The right hand begins and the left joins in with tapping eight notes in both hands. *RH-C* (right hand, complicated rhythm): the right hand begins with eight notes, and the left joins in with triplets. *LH-S* (left hand, simple rythm): the left hand starts and the right joins in with the simple rhythm (eight notes in both hands). *LH-C* (left hand, complicated rhythm): the left hand starts with eight notes, and the right joins in with triplets. (From Lang et al. 1990b)

Fig. 5. b Brain potentials in a typical subject (musical student) playing LH-S (*right*) and LH-C (*left*). Starts of LH and RH are marked by *vertical lines*. Note the BP preceding both tasks. Goniometer readings at the bottom. Note at the *right*: at the instant when RH joins in with the easy eight notes, the brain potentials show only little negativity (easy rhythm, LH-S). Note at the *left*: when RH joins in with the difficult triplets, LH has problems continuing with the eight notes (goniometer readings) and subject's SMA (lead C_Z) generates dramatic negativity with the difficult LH-C task, indicating that SMA is needed for coordinating temporally sophisticated bimanual movements. (From Deecke and Lang 1990)

have problems continuing the eight note rhythm in their LH (see goniometer readings at the bottom). At this instant their SMA (recording C_z, second from top) is dramatically tuned up, generating a very large amount of negativity for this complicated task. Thus, we think we really need the SMA for coordinating bimanual tasks, especially when they have tempo constraints, i.e., timing skills involved. Such skills are not very rare in playing music.

Figure 6A gives the overall average for the rhythm task, i.e., what we saw in Fig. 5B for a single subject, we have now as the average across the total population of our 15 musicians. RH-S is displayed by the stippled lines, RH-C by the solid lines. Negative up. We see that all subjects showed this dramatic generation of negativity (activity) starting after 4 s, particularly over SMA (C_Z), when the LH joins in with the troubling triplets (solid lines), while there is no spectacular change when the LH also taps eight notes (stippled lines).

Finally, in Fig. 6B, these potentials are topographically displayed ("mapped") using current source density analysis with spline interpolation (Lindinger et al.

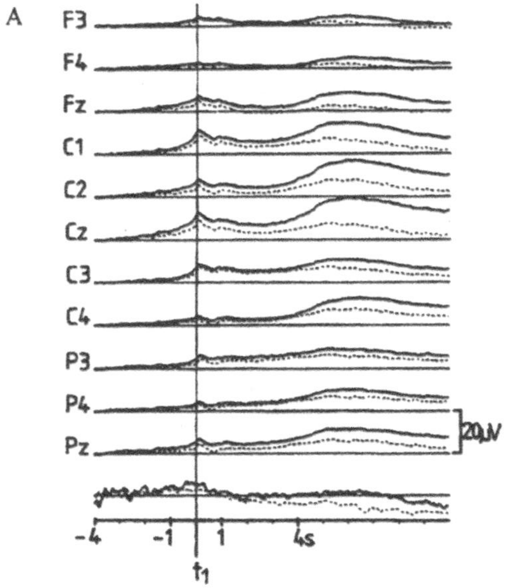

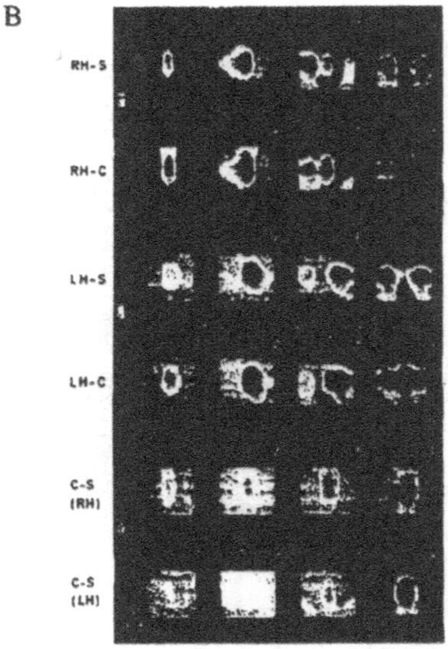

Fig. 6. a Grand average of music experiment, with the right hand (RH) leading, across all 15 subjects, time-locked to the first tap (start of RH). *Stippled line* RH-S; *solid line*, RH-C. Negative up. Note that all subjects show dramatic generation of negativity (activity) starting after 4s, particularly over SMA (C_z), when comparing *solid lines* (triplets) with *stippled lines* (eight notes).

Fig. 6. b Maps of radial current density across the scalp based on DC-potential amplitudes. *Maps 1 and 2* taken from the BP period preceding the initiation of the first tap (600-800 ms and 200-0 ms, resp. before first tap. *Map 3* while tapping 2/s eight notes in one hand (2-2.2s after first tap), and *map 4* is derived from the period of bimanual playing (8-8.2s after first tap). Note strong SMA activity (*midline*) during the complicated tasks RH-C and LH-C but not during simple tasks RH-S and LH-S. *Lower 2 rows*: Subtractions C minus S for RH and LH emphasizing the importance of the SMA′ - not the primary mptor cortex - in explaining the difference in task temporal difficulties. (From Lang et al. 1990b)

1989), i.e., the radial current density across the scalp going from positive to negative in a certain brain region is mapped. Maps are based on DC potential amplitudes. Map 1 is taken from the BP1 period 0.8-0.6 s prior to the first tap, map 2 is taken from the BP2 period 0.2-0 s prior to the first tap; map 3 displays current densities when tapping 2/s (eight notes) with one finger (2-2.2 s after the first tap); and map 4 illustrates bimanual performance (8-8.2 s after the first tap, a time at which bimanual movements were performed in all trials).The rows of maps 1 to 4 from top to bottom correspond to the tasks described in musical notation in Fig. 5A: RH-S, RH-C, LH-S, and LH-C.

(1) RH-S (top): Map 1: Already 0.8 s preceding the right hand′s start, the first activity to occur is in the SMA (oval shaped activity in the midline). Map 2: then the left MI cortex (hand area of the precentral gyrus, area 4) contralateral to the

performing RH comes into play. Map 3: Preceding the joining in of the LH, the right MI becomes active. Map 4: During the performance of the easy rhythm both motor cortices are active without SMA activation, doing the job alone, so to speak.

(2) RH-C (second row): Map 1: With the complicated task, again, the SMA leads the whole trial. However, there is already more activity - as compared to RH-S - in the SMA right from the beginning. In map 2, the left MI comes into play contralateral to the RH. In map 3, the right MI starts activity corresponding to the LH joining in, and in map 4 both MIs are active as in map 4 of RH-S. However, the great difference is in the midline (SMA): With the complicated RH-C (triplet eight notes-task), the midline is not relaxing at all as in RH-S, but is almost hyper-active, i.e., during the complicated task, the SMA is highly tuned up and generates much negativity, because it is obviously needed for such sophisticated timing jobs.

(3) LH-S is analogous to RH-S only that the left hand starts and the right comes in both with eight notes. Thus, the radial current density maps are just mirror images of RH-S: Again first SMA (map 1), then right MI (map 2), then left MI in addition (map 3) and then both MI cortices are doing the job alone (map 4).

(4) Finally LH-C is the mirror image of RH-C and has the same strong SMA activation during the performance of the complicated quaver-triplet-rhythm.

In the lower two rows of Fig 6B, corresponding complicated and simple rhythm tasks have been subtracted (fifth row, C - S (RH): RH-C minus RH-S and sixth row, C - S (LH): LH-C minus LH-S). We see that the MI activities are totally subtracted since they are the same in simple and complicated tasks. The difference is in the SMA! This is so highly more tuned up during the C tasks than in the S tasks, in that it shows this significant extra negativity. SMA (not MI) is the critical structure, when differences in *temporal* task difficulties are under study.

Conclusion

We have seen that using the DC potential method, fundamental insights into human voluntary movement organization can be gained. As examples general voluntary action and its brain potentials has been described as well as the potentials related to motor learning, potentials with SMA damage, and the potentials accompanying rhythmic coordination of the hands in playing music. Especially the latter can tell us that the SMA is highly important for controlling the sophisticated bimanual timing skills that are the essence of instrumental musical performance. If we even consider an organist or a percussionist, it is not just bimanual but rather bipedal coordination that is needed. Such precise temporal coordination and integration of the two halves of the body surely requires not only intact but particularly well developed and trained SMAs.

References

Asenbaum S, Oldenkott B, Lang W, Lindinger G, Deecke L (1990) Motor sequences in patients with unilateral SMA lesions: movement-related potentials and motor disorders. In: Brunia CHM, Gaillard AWK, Kok A (ed) Psychophysiological brain research, Vol. 1, Tilburg University Press, Tilburg pp 119-123

Boschert J, Hink RF, Deecke L: Finger movement versus toe movement-related potentials: further evidence for supplementary motor area (SMA) participation prior to voluntary action. Exp Brain Res 55:73-80

Deecke L (1987) Bereitschaftspotential as an indicator of movement preparation in supplementary motor area and motor cortex. In: Porter R (Chair) Motor areas of the cerebral cortex. Ciba Found Symp 132:231-250

Deecke L., Kornhuber HH (1978) An electrical sign of participation of the mesial "supplementary" motor cortex in human voluntary finger movements. Brain Res 159: 473-476

Deecke L, Lang W (1989) Motorische Hirnrindenfelder und rhythmische Koordination der Hände beim Musizieren. In: Petsche H (ed) Musik - Gehirn - Spiel. Beiträge zum 4. Herbert von Karajan-Symposium, Birkhäuser, Basel, pp 163-178

Deecke L, Lang W (1990) Movement-related potentials and complex actions: coordinating role of the supplementary motor area. In: Eccles JC, Creutzfeldt O (ed) The principles of design and operation of the brain. Pontificia Academia Scientiarum Scripta Varia 78: 303-341

Deecke L, Lang W, Heller HJ, Hufnagl M, Kornhuber HH (1987) Bereitschaftspotential in patients with unilateral lesions of the supplementary motor area. J Neurol Neurosurg Psychiatr 50:1430-1434

Deecke L, Lang W, Uhl F, Podreka I (1990)Looking where the action is: negative DC-shifts as indicators of cortical activity. In: Deecke L, Eccles JC, Mountcastle VB (ed) From neuron to action: an appraisal of fundamental and clinical research. Springer Berlin, Heidelberg, New York pp 25-41

Foerster O (1934) Motorische Felder und Bahnen. In: Bumke O, Foerster O (ed) Handbuch der Neurologie, IV. Springer, Berlin, pp 1-448

Goldenberg G, Lang W, Podreka I, Deecke L (1990)Are cognitive deficits in Parkinson´s disease caused by frontal lobe dysfunction? J Psychophysiol 4: 137-144

Kornhuber HH, Deecke L (1964) Hirnpotentialänderungen beim Menschen vor und nach Willkürbewegungen, dargestellt mit Magnetbandspeicherung und Rückwärtsanalyse. Pflugers Arch Ges. Physiol 281:52

Kornhuber HH, Deecke L (1965) Hirnpotentialänderungen bei Willkürbewegungen und passiven Bewegungen des Menschen: Bereitschaftspotential und reafferente Potentiale. Pflugers Arch Ges Physiol 284: 1-17

Kristeva R, Deecke L (1980) Cerebral potentials preceding right and left unilateral and bilateral finger movements in sinistrals. In: Kornhuber HH, Deecke L (ed) Motivation, motor and sensory processes of the brain: electrical potentials, behaviour and clinical use. Prog Brain Res 54:748-754

Lang W, Lang M, Kornhuber A, Deecke L, Kornhuber HH (1983) Human cerebral potentials and visuomotor learning. Pflugers Arch 399:342-344

Lang W, Lang M, Kornhuber A, Kornhuber HH (1986) Electrophysiological evidence for right frontal lobe dominance in spatial visuomotor learning. Arch Ital Biol 124:1-13

Lang W, Lang M, Podreka I, Steiner M, Uhl F, Suess E, Müller C, Deecke L (1988a) DC-potential shifts and regional cerebral blood flow reveal frontal cortex involvement in human visuomotor learning. Exp Brain Res 71:353-364

Lang W, Zilch O, Koska C, Lindinger G, Deecke L (1988b) Negative cortical DC shifts preceding and accompanying simple and complex sequential movements. Exp Brain Res 74:99-104

Lang W, Goldenberg G, Podreka I, Cheyne D, Deecke L (1990a) Parkinsonism as a disturbance of movement initiation. J Psychophysiol 4:123-136

Lang W, Obrig H, Lindinger G, Cheyne D, Deecke L (1990b) Supplementary motor area activation while tapping bimanually different rhythms in musicians. Exp Brain Res 79:504-514

Lang W, Cheyne D, Kristeva R, Beisteiner R, Lindinger G, Deecke L (1991) Three-dimensional localization of SMA activity preceding voluntary movement. A study of electric and magnetic fields in a patient with infarction of the right supplementary motor area. Exp Brain Res 87:688-695

Lassen NA, Ingvar DH, Skinhoj E (1978) Brain function and blood flow. Sci Amer 239:50-59

Lindinger G, Lang W, Deecke L (1989) Topographische Analysen bewegungskorrelierter, kortikaler DC-Potentiale mit Hilfe von Spline-Interpolation und Stromquellendichteanalyse. Z EEG-EMG 20:122

Penfield W, Rasmussen AI (1950) Cerebral cortex of man. A clinical study of localization of function. Mac Millan, New York

Vogt C, Vogt O (1919) Allgemeinere Ergebnisse unserer Hirnforschung IV. Die physiologische Bedeutung der architektonischen Rindenfelderung auf Grund neuer Rindenreizungen. J Psychol Neurol (Leipzig) 25:339

DC-EEG, MEG and FMRI as Investigational Tools for Music Processing

ROLAND BEISTEINER

Introduction

In the past decades there has been considerable progress in the development of new tools for monitoring specific brain activity. With these it is possible to address a wide variety of cognitive and motoric tasks and thus questions such as: *What brain areas are active when listening to music? What brain areas are responsible for performing different kinds of processing of music? How is the brain organized when performing complex motoric behavior like necessary to play an instrument?*

Three of the latest powerful tools will be introduced here, and their usefulness in the investigation of musical processes discussed.

First, in the 1960s, increasing interest developed for the use of slow potential recordings (direct current electroencephalography DC-EEG) in EEG research (Kornhuber and Deecke 1965). For true DC - recordings high-tech amplifiers with an infinite time constant are needed. Such amplifiers allow for long-term recordings (up to hours) without artificial drifts of the recording baseline.

A second method, very closely related to EEG recordings, started to be used in the 1980s for clinical research - the magnetoencephalography (MEG, see Deecke et al. 1982). MEG measures the magnetic fields associated with the electric currents detected by EEG. Since the magnetic resistences of brain, body fluids, bone and skin are of comparable size whereas the electric resistences are not, there will be considerably less distortion in the magnetic fields than in the electric fields on their way from the brain source to the detector. As a consequence it can be assumed that the localization of active brain sources will be more accurate with MEG measurements.

Currently, so to speak in the 1990s, a promising new tool is about to develop: functional magnetic resonance imaging (FMRI, Belliveau et al. 1991) for the first time will allow simultaneous production of a highresolution picture and a functional map of the living brain. FMRI is based upon changes in signal intensity with changes in blood volume and blood velocity, considering the fact that local activity increases local blood supply. Unlike with other methods, measurements can be taken without any radioactive contrast agent and as such, without any danger for the subject.

To demonstrate the applicability of the different methods for research in music, three experiments, conducted in Vienna and concerning music specific activity, will be presented here.

Methods

Direct Current Eletroencephalography

DC-EEG allows measurement of ongoing electrical activity of the brain when performing tasks. The electrical activity measured on the scalp reflects activity of the underlying cortex in the sense that there is a negative potential shift when the underlying cortex is activated. We used 11 different locations on the scalp to look for these task-related negativities (F3, F4, F7, F8, T3, T4, T5, T6, P3, P4, Oz). The purpose of our DC-EEG study was to investigate analytic, creative, and mnemonic brain processes while listening to and processing music. Students of music listened to a sequence of four tones and subsequently either were to reverse the sequence [task (1) = analytic] or to compose a new continuation [task (2) = creative]. In task (3), the initial segment of a well-known melody was presented and had to be continued (mnemonic task). All tasks had to be done in the mind (imagery). In (1) and (2) either tonal or atonal sequences were presented.

Magnetoencephalography

Using MEG the magnetic fields associated with the electric currents in active brain regions can be measured. We present the preliminary results of a study addressing the question of whether it is possible to observe what is happening in the brain when analyzing harmony. To achieve this goal we presented the first three chords of an ordinary musical cadence - I, IV, V - but instead of the I as the fourth chord we presented a target tone, which was either a harmonically fitting tone (in 75% of the cases either the fundamental, the third, or the sixth of the tonic) or a harmonically nonfitting tone (in 25% of the cases the minor second). The nonfitting target had to be counted. Each of the four target tones occurred with equal probability (tonic, third, sixth, minor second) and in total we used 11 different pitches (completely randomized). All stimuli were presented to the left ear. Five to seven measurement positions over the right temporal area were used to record the magnetic fields associated with processing of the target tones. The indicator used for observing tone processing was the magnetic P300 (P300m). In electrophysiology P300 means a positive wave 300 ms after onset of an infrequently occurring stimulus to which the subject is attending. It indicates the termination of some aspects of cognitive stimulus processing such as determining whether the target tone belongs to the harmonic context heard before or not. By using the same tones in different harmonic functions for the fitting and for the nonfitting targets we made sure that a differentiation could only be performed by refering to the harmonic context heard before. In addition we tried to minimize memory effects regarding the test stimulus by using tones instead of chords (otherwise a differentiation could have been based on the similarity of the first chord with the test chord). If enough data points are taken from a large area of the skull, it is possible to find the magnetic field maxima and minima associated with the P300m. Given that, it is possible to reliably localize the center of the P300m activity within the human brain.

Functional Magnetic Resonance Imaging

FMRI is a promising new tool for brain imaging, currently under development. FMRI allows for simultaneous production of pictures and functional maps of the brain without any danger or harm to the subject. The method is based upon conventional magnetic resonance imaging (MRI), in which electromagnetic signals of stimulated atoms are used to generate pictures of human tissue. It is possible to select and observe any picture slice through the human corpus wanted. The basic hypothesis for functional imaging is that the deoxyhemoglobin of blood is a paramagnetic contrast agent which reduces local signal intensitiy. Active brain regions show a great increase of local blood volume and a large inflow of non-field-attenuating diamagnetic oxyhemoglobin, such that the concentration of the signal-attenuating deoxyhemoglobin decreases, resulting in an increase of signal intensity from the active areas. The nonactive areas show no blood volume change and no such signal change. In addition, there is good reason to assume that blood velocity changes alone may account for some of the observed effects (Gomiscek et al. 1993).

We present the preliminary results of a pilot study in which subjects performed complex movements with wrist and hand, designed to mimic partly the complexity of motoric processes demanded of instrumentalists. The movements were as follows: extending the right wrist, extending the left wrist, extending the right index finger, extending the left index finger, flexing the right wrist, flexing the left wrist, flexing the right index finger, flexing the left index finger, and repeat again without pausing. The regions of interest comprise the primary sensori-motor cortex and parts of the supplementary motoric area (SMA). The subject alternates 10 periods of 2 min moving and 10 periods of 2 min resting. The difference in the pictures of moving and resting show the brain regions activated with the task.

Results and Discussion

Direct Current Electroencephalography (Fig.1)

The creative task showed the lowest overall activity and thus the other tasks were contrasted against it. Comparing the analytic with the creative task, the first showed

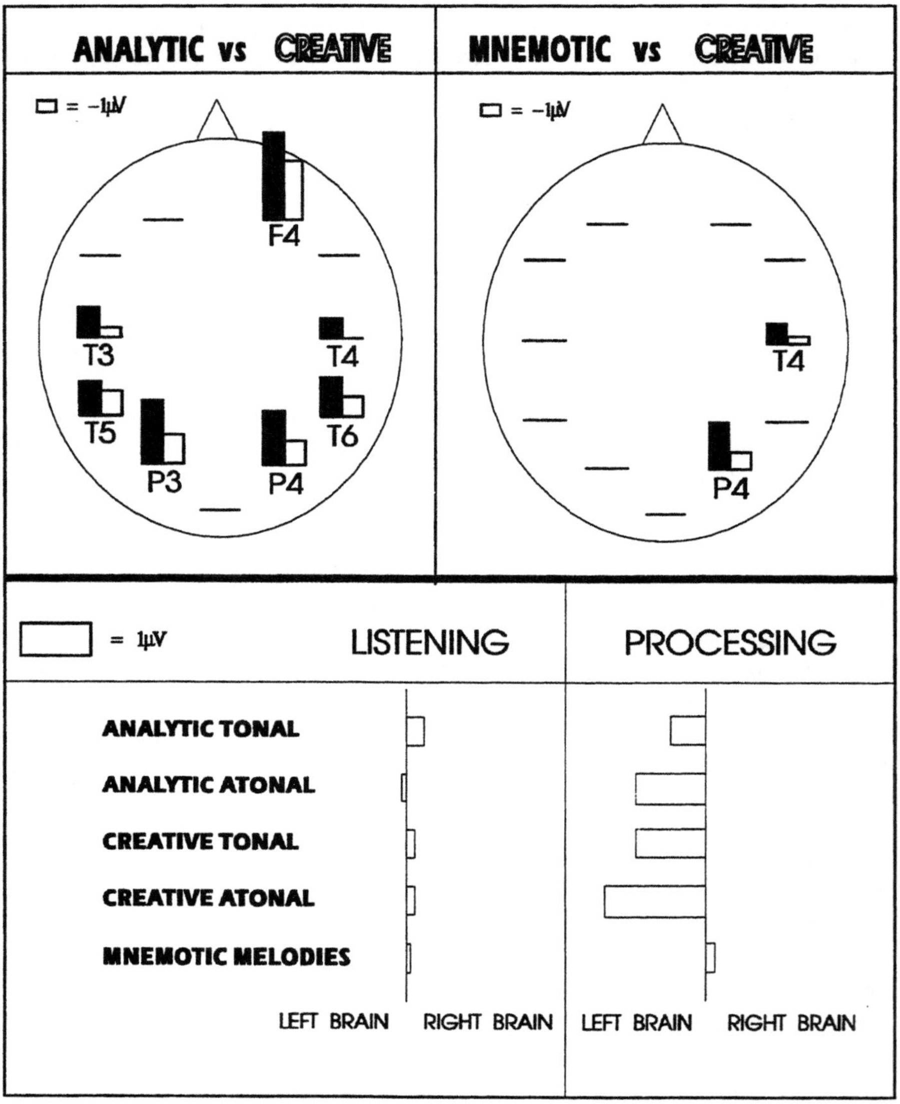

Fig. 1. Results of the direct current electroencephalography experiment. The *upper half* presents electrodes which showed significant differences between the named tasks. Note that with the mnemonic task there is an amplitude increase only over the right hemisphere.
The *lower half* presents task-specific lateralization tendencies. Note the general left shift when processing except with the mnemonic task, where the right-hemispheric tendency increases. The only significant lateralization effect is the change from right to left in the creative task

increased involvement of the parietotemporal regions of both hemispheres. The mnemonic task, however, showed only increased activity over the right hemisphere. Lateralization results confirmed the significance of the right hemisphere for the mnemonic task, since this was the only one showing a right-hemispheric tendency when both listening and processing of music. The other tasks showed a general tendency to the right hemisphere while listening and to the left hemisphere while processing. Surprisingly, this finding was strongest with the creative task (showing the only *significant* lateralization effect). Creativity was expected to be a right hemispheric, wholistic-synthetic task und we thus expected a right-hemispheric preponderance. An explanation for the left-hemispheric involvement with composing may be that our task possibly comprised important analytical demands such as self-evaluation of the outcome according to given external or internal normative patterns. Since the present study is the first to provide electrophysiological correlates of creativity in music, further research must be conducted before a definite interpretation of left predominance with creativity is possible.

According to our results, it may be concluded that activity in musicians does not per se lateralize to the left hemisphere when listening to or processing music. It probably depends on the amount of simultaneous analytic - sequential processing whether it is possible to find a left-hemispheric preponderance with musicians.

Magnetoencephalography (Fig.2)

The main question of this study was whether it is possible to find correlates of harmonic processing in the human brain. If the subject was able to differentiate

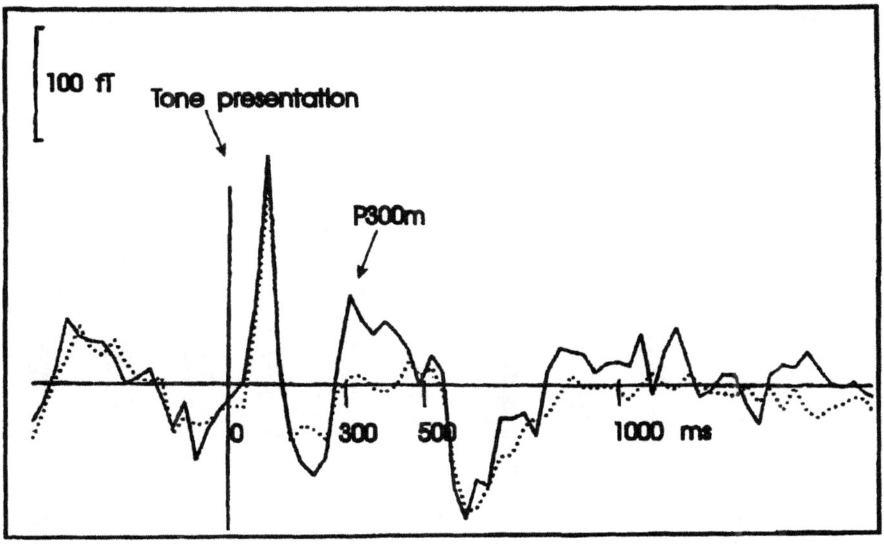

Fig.2. Magnetoencephalography waveforms as recorded at one channel. *Solid line* shows magnetic field course following the minor second. Note the P300m starting about 300ms after tone presentation. *Dotted line* shows the tonic fundamental

whether the target tone fits the preceding harmonic context or not, a distinct magnetic wave starting around 300 ms post-stimulus was expected (P300m) to occur in the nonfitting condition (minor second). Note that a differentiation of the "harmonic load" of the target tone could only be performed as a consequence of the preceding chords since every target tone was preceded by harmonically fitting as well as non fitting contexts. Results show that subjects who are able to perform well indeed show such a P300m. Thus further development of this design might provide the possibility for objective testing of harmonic listening capabilities of humans.

Functional Magnetic Resonance Imaging (Fig.3)

Preliminary results of our Viennese group (comprised of members of the Neurological University Clinic, the Institute of Medical Physics, and the Institute of Magnetic Resonance) show the feasibility of monitoring brain activity and simultaneously acquiring pictures of the monitored region. Subjects had to perform a complex motoric task and this resulted in an increase of signal intensity in brain regions comprising the hand region of the primary motor cortex and parts of the supplementary motoric area. The subjects alternately had to move or to pause and two functional pictures were acquired within every period. In total 10 moving pictures and 10 resting pictures were paired according to the sequence of events. Thus 10 moving - resting differences could be calculated and resulted in a signal increase with moving in eight of 10 cases in the physiologically expected region. The method has to be improved especially concerning movement artifacts.

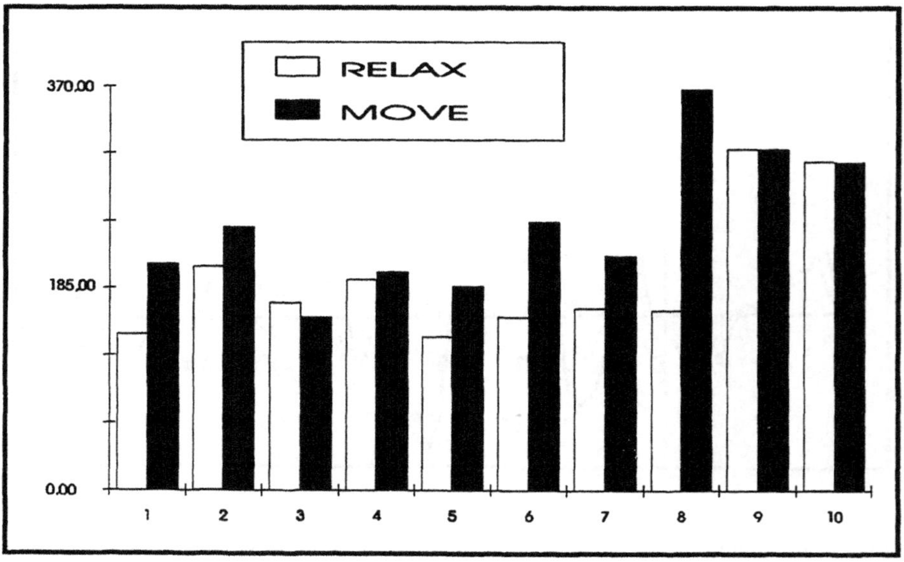

Fig.3. Signal intensities measured in the functional magnetic resonance imaging experiment. Intensities for RELAX (*open bars*) and MOVE (*black bars*) are shown

Concerning the general applicability of all three methods introduced, it should be noted that the main difference concerns the time resolution. EEG and MEG are incomparabel in time resolutions down to milliseconds. The lowest expected time resolution with FMRI, will probably not be much below 1 s. The reason for this is that FMRI is dependent on blood volume changes accompanying brain activity, which start at a latency of about 1 s and reach maximum after about 3-10 s. However, FMRI is incomparable in spatial resolution. Thus, combining EEG, MEG and FMRI means combining high temporal with high spatial resolution and will open new horizons for cognitive music research.

References

Belliveau JW, Kennedy DN, McKinstry RC, Buchbinder BR, Weiskoff RM, Cohen MS, Vevea JM, Brady TJ, Roasen BR (1991) Functional mapping of the human visual cortex by magnetic resonance imaging. Science 254: 716-719

Deecke L, Weinberg H, Brickett P (1982) Magnetic fields of the human brain accompanying voluntary movement: Bereitschaftsmagnetfeld. Exp Brain Res 48: 144-148

Gomiscek G, Beisteiner R, Hittmair K, Müller E, Moser E (1993) A possible role of in-flow effects in functional MR-imaging. Magn Reson Mater 1:109-113

Kornhuber HH, Deecke L (1965) Hirnpotentialänderungen bei Willkürbewegungen und passiven Bewegungen des Menschen: Bereitschaftspotential und reafferente Potentiale. Pflugers Archiv 284: 1-17

Music and SPECT

Sören Nielzén

The perception and expression of music involves activities at high levels of functioning in the human brain. This requires imagination, associations, focused attention, learning, deliberate contemplation, and other things connected with cognitive processes. An important aspect of music is its appeal to bodily and emotional experiences by rhythm, color, timbre, and other components. Thus most fields of neurophysiological functioning are activated by music.

In order to understand the demands and frames of music in communication, execution, composition, and so forth, music has to be viewed in relation to neurophysiology, psychology, and other human sciences. Useful tools for neurocorrelation studies have been offered during the last 20 years by various neuroimaging techniques, and one of them, the Single photon emission computerized tomography (SPECT)-technique, will be reviewed in more detail in the following.

The SPECT technique, developed from early methods of measuring the cerebral blood flow using radioactive tracers (Lassen 1987). This can be accomplished by injecting these tracers into an artery, a vein, or by inhaling them. The radiation is then recorded with a great number (40-250) of detectors placed directly over the skull. From the pattern of pulses registered, a map of the regional blood flow in the underlying layer of cortex may be derived. The SPECT system constitutes an amelioration of this method in the sense that it permits the construction of three-dimensional (3-D) images by its more sophisticated system for detection of radiation. Events at sites even under the cortex may be followed.

Single photon emission computerized tomography, SPECT, is an expression that fairly well describes the physical realities of the imaging technique. A fat-soluble -radiant agent is inhaled or injected intravenously into the subject being investigated; this agent is transported to the brain tissue by the general circulation of blood. There it is distributed according to the present pattern of neural activation, since the activity of the neural tissue controls the regional cerebral bloodflow in all parts of the brain. When the activity is low, the bloodflow is decreased, and when it is high the bloodflow increases. In the SPECTmachine (Tomomatic 564 and others; Medimatic A/S, Copenhagen, Denmark), crystals arranged in a rotating circular collimator transform the -radiation into light. Signals of photons are produced which are caught by peripherally located detectors at some distance from the collimator. In a later step the data obtained regarding direction and timing of the pulses, the computer creates images, "back-projects" the data into a 3-Dl image of the original distribution of increased or decreased activity. It is also possible to construct images of certain parts of the brain or of slices in different planes. These pictures have a resolution of 8-20 mm, depending on isotope and camera technique, which means that not very small areas can be studied. For exact orientation and for interpretation more precise anatomical registrations with, for example, magnetic resonance tomography (MRT) or computed tomography (CT) may be used as well.

The limitations of the SPECT method are equally important to consider in the planning of experiments. If the radioactive isotope ^{133}Xe is used, it is possible to calculate an absolute value of the rate of bloodflow at rest (50ml/100 g tissue per minute on average). The recording takes about 5 min, but may be repeated within half an hour. Changes of treatment or new stimulations may be compared in this way. In other words the tracer can be followed dynamically. ^{99m}Tc ligated to a carrier substance such as hexamethyl-propylene-amine-oxime (HMPAO; a compound commercially available under the name of Ceretec) is most commonly used as tracer nowadays. This substance becomes incorporated for a longer time in the nerve cells, and measurements can be undertaken over several hours.

But the resulting image only reflects the initial state of activation, and the relative distribution is shown without any possibility of calculating absolute flowvalues. It is also theoretically possible to use specific ligands other than HMPAO and trace processes related to for example specific transmitter systems of the brain (Konig et al. 1991). This extended use of the method has not yet become common, partly due to economic considerations, and partly to the existence of another technique widely used for such purposes, positron emission tomography, or PET.

The SPECT technique has advantage over certain other brain-imaging techniques due to the relative cheapness of each procedure and the low radiation intensity the subjects are exposed to. This is 5 mSi, which is not more than what one gets in the natural environment in 1 year in Sweden.

The SPECT- technique is most commonly used in the medical examination of neurological diseases such as dementia and cerebrovascular disease but also to study functional differences in psychiatric and in psychological conditions. Neurophysiological studies with sensory stimulation and SPECT have taught us much about the brain´s organization of normal perception and behavior. It has also revealed some major differences from normal functional patterns in psychopathological states, as, for example in schizophrenia.

Studies with special relevance to musical perception and performance are rare, however, a few have been performed in recent years. In a study by Ryding et al. (1987), the regional rate of cerebral blood flow rCBF was measured while humming (without words) a simple and well-known nursery rhyme and during repetitive recital of the weekdays in order to investigate speech function. Automatic speech was found to involve the left hemisphere, regarding the activation of Broca's area and the precentral motor regions, i.e., speech motor preparation programs and the motor control of the mouth and tongue. The right hemisphere was involved mainly in laryngeal motor control as well as in the feedback of reception of sensory and auditory information as indicated by the rolandic and temporal activations, respectively. While humming, the right hemisphere showed focuses of activation in the rolandic area, in the temporal area, and in the frontal area involved in the recognition of auditory information, voluntary control, and memory activation.

The left hemisphere showed a pattern of rCBF, indicating that it took part in the voluntary control, the preparation of motor programs for the hummed tune, and in memory activation. The authors conclude that speech function and musical function are not localized only to the left hemisphere, as originally proposed by Broca and Wernicke. Instead they point out that speech and musical functions require collaboration of both hemispheres in various functional patterns.

An interesting further step towards the demonstration of such functional patterns was presented by McLaughlin et al in 1992. The methodological innovation in this study was to intercorrelate the values of rCBF of various regions and in this way connect different regions which in the same time periods work together. Correlations were computed for three different conditions of stimulation: A diotic stimulation consisting of the narration of a Danish narrative, a dichotic stimulation of 12 wordpairs, and a presentation of white noise between 500 and 10 000 Hz. The subjects were asked to memorize the narrative and the wordpairs. Then, a factor analysis was performed, which reduced 28 regions of interest to a set of three independent factors accounting for 77% of the total variation in rCBF values. In other words three systems or networks in the brain were discovered that were involved in the analysis of language. The authors assigned the functions of these networks to be one of auditory/linguistic interpretation, one of attention, and one of visual imagination. Several advantages of the method used in this study are discussed. The approach helps to avoid the need for artificial experimental paradigms, e.g., subtracting silent counting from overt speech in order to obtain a "pure" function. The use of factor analysis further reduces redundant information from multiple brain regions into a set of independent factors. The risk of the use of overall t-testing which may result in false-positive findings is also reduced.

Another field of interest which touches upon music and auditory perception is the auditory hallucination. This has been the subject of several SPECTstudies, one of which will be reviewed here. Walter et al.(1990) induced hallucinations in 10 normal subjects by hypnotic suggestion and managed to induce musical hallucinations. Originally they aimed at initiating speech hallucinations, which, however was not successful. Therefore, their study compares musical hallucinations in these normal subjects with verbal hallucinations of varying causes in 10 psychiatric patients. The activation of different regions of the brain during ongoing hallucinations was then studied by SPECT using HMPAO as a tracer. Four slices, cranial to caudal, of the brain were used for comparison in the two groups of subjects. In the first slice the normal subjects showed less activity in the left frontal and in the left central region. In the second slice the hypnosis group showed an increase of rCBF in the left and right superior frontal region and in the left mesio-frontal region. There were also generally higher levels of rCBF in both hemispheres. In the third slice there was higher activity during music hallucinations in thalamic regions connected with higher values in the right hemisphere. The fourth slice revealed lower activity in in mesio-temporo-limbic and hippocampal structures in the hypnosis group than in the patient group . This study again emphasizes the brain's engagement of various net-works depending on the type of auditory processing going on.

It was postulated in the introduction that musical activities engage very complex mental processes. Some proof of this has now been presented, more or less indirectly. An interesting study from Canada dealing exclusively with musical activities and brain responses was published in Science (Sergent et al. 1992). The aim of the study was to aquire some knowledge of what is going on while playing music on a keyboard, which can be supposed to activate different sensory areas of visual and auditory functions and motoric areas. This time recordings were made for rCBF measurements. This is another brain-imaging technique, for which additional tracers are available. However, only rCBFmeasurements were used in the study, which means that, from a theoretical standpoint, it is similar to SPECTinvestigations, and indeed the same rCBFmeasurements could have been performed with SPECT. For this reason it may be interesting to look at it in this very context, where music is the main theme of interest.

Ten professional pianists were used as subjects. They were asked to read, listen to, and play with their right hand a melody unknown to them from a "Partita" by J.S.Bach. A number of controlling procedures were undertaken first, such as sightreading a score without playing, playing up and down scales, and listening to a melody from another unknown "Partita" by the same composer. Values from measurements of these preparatory tasks were then subtracted from values of the main task in order to obtain an image more precisely related to the main task.

The preparatory tasks revealed different patterns of cortical activity consistent with other findings with similar techniques on auditory perception and motor performance. Furthermore, new areas of activity related to the main task were discovered, that were not involved by playing, listening, and reading alone. One area involved the superior parietal lobe of both hemispheres, reflecting the generation of spatial information derived from the location of the notes on the staff for the actual motor performance required in the manual execution of the musical piece. In other studies this area has been shown to mediate sensorimotor transformation for visually guided skilled actions and finger positioning. The other area reflecting processes of the main task was the left premotor cortex and the left inferior frontal gyrus immediately above Broca's area reflecting the motor sequencing required for the keyboard performance.

In summary, the task of playing a melody was shown to evoke a cerebral network distributed over the four lobes and the cerebellum. This network parallels the neural substrate of verbal processing, but yet is distinct from it. The localizations are in the proximity of regions demonstrated earlier as being engaged in auditory and motor tasks. Furthermore, melody playing requires the participation of the superior parietal lobe for the spatial processing of reading the music. Besides this, the study contributes to the explanation of why so many uncorrelated, different clinical patterns of agnosia, amusia, aphasia, agraphia, etc. exist.

Even if SPECT methods have yielded most substantial contributions to the understanding of auditory perception and musical performance, we are far from understanding the experiences and reactions elicited by music. When investigating the effect of musical stimulation and the neurophysiological processes involved in musical performance, we still have to search for more exact and additional means measuring physiological processes. We must deepen our knowledge of the componential structuring of music as a stimulus and we have to further analyze the specific mental mode of functioning within the domain of music.

References

Konig P, Benzer MK, Fritzsche H (1991) SPECT-technique for visualization of cerebral dopamine D2 receptors. Am J Psychiatry 148:11

Lassen NA (1987) Cerebral blood flow measured by Xenon-133. Nucl Med Commun 8:535-548

McLaughlin T, Steinberg B, Christensen B, Law I, Parvin A, Friberg L (1992) Potential language and attentional networks revealed through factor analysis of rCBF data measured with SPECT. J Cereb Blood Flow Metab 12: 535-545 5

Ryding E, Brådvik B, Ingvar DH (1987) Changes of regional cerebral blood flow measured simultaneously in the right and left hemisphere during automatic speech and humming. Brain 110:1345-1358

Sergent J, Zuck E, Terriah S, McDonald B (1992) Distributed neural network underlying musical sight-reading and keyboard performance. Science 257:106-109

Walter H, Prodreka I, Steiner M, Suess E, Benda N, Hajji M, Lesch OM, Musalek M, Passweg V (1990) A contribution to classification of hallucinations. Psychopathology 23:97-105

EEG Mapping During Music Perception and MRI Data in Psychiatric Patients

REINHARD STEINBERG, WILFRIED GÜNTHER, UTE KLAGES, CHRISTIAN HAAG, THOMAS VOGL AND PETER STRECK

Introduction

Performance and perception of music are reversibly disturbed in mental disease as are cognition, affect and psychomotor behavior (Steinberg et al. 1985, 1989; Günther et al. 1991). With the beginnings of psychiatry the search for morphologic correlates of alienation and disturbed mood accompanies all scientific efforts, since an organic basis for at least a considerable part of those illnesses is strongly assumed. Computer-based brain imaging technologies which have emerged in the last two decades have broadened considerably the accessibility to morphology and also to in vivo functionality of the nervous system, the latter being even more and more important in the understanding of pathophysiologic mechanisms underlying psychotic experience and behavior. In schizophrenia computed tomography (CT) and magnetic resonance imaging (MRI) often reveal atrophy of limbic structures and prefrontal areas in patients who underwent a change in personality after a chronic course of the disease. Microscopically, a tilted position of cortical neurons in limbic structures is found, a pecularity which assumably develops in the second trimester of gestation. It suggests an early disturbance caused by genetic and maybe additional environmental factors (Beckmann et al. 1987).

Functional imaging with positron emission tomography (PET) and single photon emission tomography (SPECT) suggest with these type II patients (Crow 1985) considerably lower glucose utilization and abnormal flattening in regional cerebral blood flow under usual stimulation conditions. Otherwise, type I patients with "positive" psychotic symptoms as, for example, vivid delusions, hallucinations, or elated or depressed mood seem to lack atrophic processes. Corpus callosum, which connects the two hemispheres, seems to be even bigger in diameter compared to controls (Günther et al. 1989). Also cerebral blood flow rises considerably higher in some cortical and subcortical areas than in matched controls and type II patients. The involvement of dopamine metabolism in the pathophysiology of schizophrenia could be further elucidated, and the different distribution of the pathways and its receptors partially confirmed in vivo as well as modified in favor of other mechanisms involved, such as the glutaminergic system. But up to now those examinations are restricted to a few centers because it requires a lot of time, personnel and money and inevitably involves the use of radioactive tracers as well.

Computerized EEG techniques contribute to the functional examination of the cerebrum. The disadvantages of a rather low local resolving power and restriction mainly to cortical events are balanced by low costs, a comparably unlimited applicability, which also enables long-term studies to be conducted without burdening the subjects too much, and an excact temporal resolving power. After fast Fourier transformation the power of the genuine EEG or especially triggered fast components of evoked potentials can be computerized and mapped according to the electrode positions on the scalp.

The variety of methods allows us to access the brain and its function in a variety of ways, but at the same time raising the question of comparability and validity of the findings. Here preliminary results of an extensive collaborative study are reported, in which EEG mapping data are cross-validated with the results of other morphologic and functional studies of the brain in the course of treatment of mental patients. Here, the preliminary results of EEG mapping and MRI examinations with schizophrenic patients and matched controls are discussed.

Method

A 32-channel mapper (Brainstar/Schwind, Erlangen, Germany) was used according to the 10/20 system. Thirty EEG channels and two EOG channels were recorded with two connected ear electrodes as reference. Relexation for 1 min with eyes closed between each stimulation served as reference state. Stimulation consisted of repetitive closing of the right fist (manumotoricity) and two musical stimuli, presented over earphones for 1 min. Music 1 consisted of a rhumba rhythm, presented as a repetitive arpeggio in D-major with MM 120, generated automatically by a keyboard. Music 2 presented the first minute of the first movement (allegro) of Mozart´s Hunting Quartett (KV 377). The EEG was carefully controlled for artifacts. Periods longer than 6 s were taken for analysis and relative power defined within the common frequency bands delta (0.5-3.5 Hz), theta (4-7.5 Hz), alpha (8-13.5 Hz) and beta (14-30 Hz; for details see Günther et al. 1991).

MRI examination was conducted with a 1.5 Tesla Siemens Magnetom, 256x256 matrix, T2-weighted scans. Nineteen linear parameters concerning possible atrophy within the brain were defined.

Patients underwent EEG mapping examination shortly after admission before receiving neuroleptic medication. Pretreated patients were examined after a wash out period of at least 2 weeks. Informed consent was received according to the Helsinki criteria.

Results

As is shown in Table 1, 42 schizophrenic patients and 61 controls underwent EEG mapping examination. With a mean age of 38 years the female schizophrenics in

Table 1 Mean age and psychopathologic status of 42 schizophrenic patients and 61 controls

	$\bar{x}$	SD
Schizophrenic patients (*n* = 42)		
Age	31.8	11.5
ANDP	13.4	3.8
ANER	12.9	4.7
THOT	13.0	4.8
ACT	7.8	3.8
HOST	9.8	4.3
BPRS	56.9	12.4
SANS	65.8	28.6
SAPS	52.1	21.5
Male schizophrenic (n = 21)		
Age	25.7	5.6
ANDP	12.9	3.9
ANER	12.4	4.7
THOT	13.1	4.6
ACT	8.3	3.8
HOST	9.1	4.1
BPRS	55.8	11.7
SANS	66.9	31.6
SAPS	53.7	24.6
GAS	35.2	10.9
Female schizophrenics (n = 21)		
Age	37.9	12.7
ANDP	13.9	3.8
ANER	13.4	4.7
THOT	12.8	5.2
ACT	7.3	3.9
HOST	10.5	4.4
BPRS	57.9	13.3
SANS	64.8	26.1
SAPS	50.6	18.3
GAS	33.8	8.2
Controls (n = 61)		
Age	33.5	12.1
Male controls (n = 32)		
Age	32.0	11.3
Female controls (n = 29)		
Age	35.2	12.9

Fig.1 a

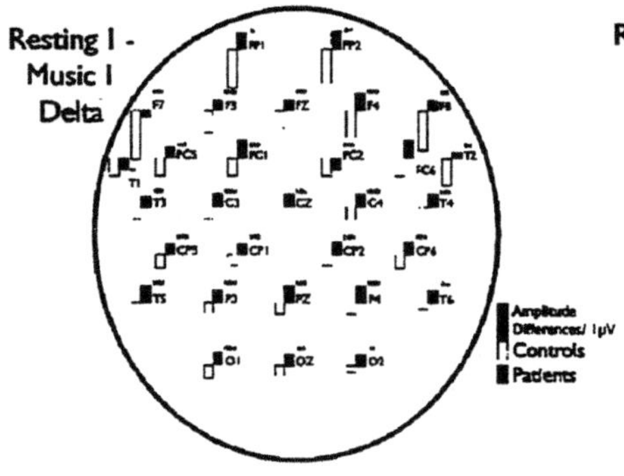

Fig.1 b

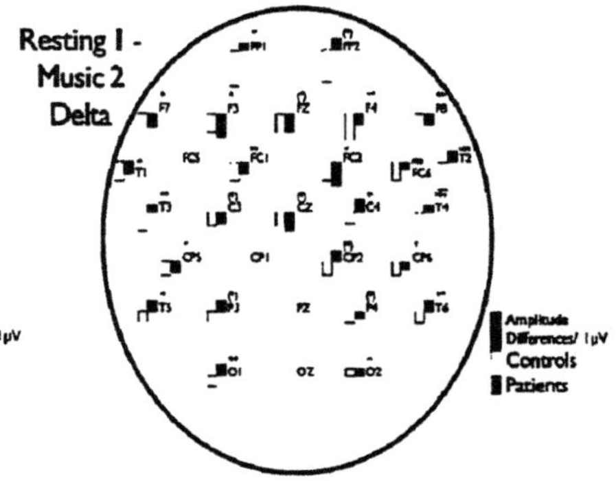

Fig.1 c

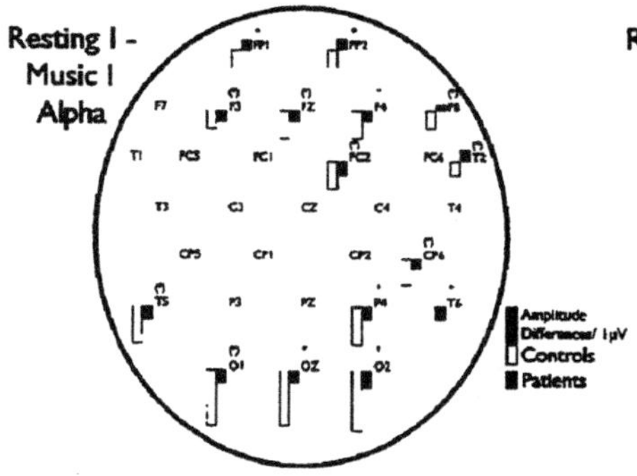

Fig.1 d

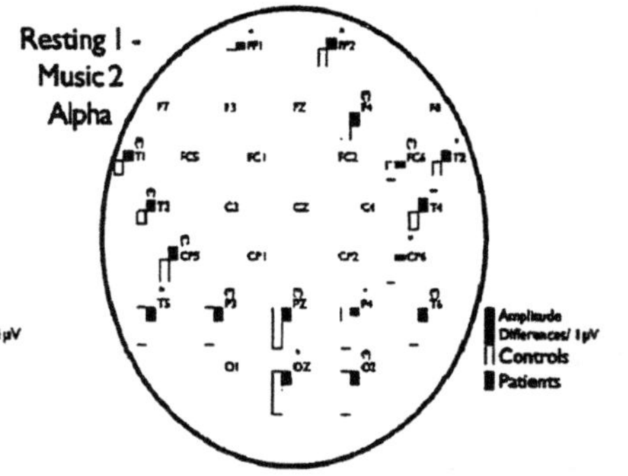

Fig.1 a–d EEG power changes during music stimulation compared to resting conditions. Electrodes placed according to the international 10/20 nomenclature. (*) $p < 0.1$; * $p < 0.05$; ** $p < 0.01$; *** $p < 0.001$) **a** Delta power with music I stimulus. **b** Delta power with music II. **c** Alpha-Power with music I. **d** Alpha power with music II.

the sample were considerably older than the male schizophrenics. Within the control group there was not as much age difference between genders. Otherwise, the Brief Psychiatric Rating Scale (BPRS; Overall and Gorham 1962) scores and the factors anxiety/depression (ANDP), anergia (ANER), thought disorders (THOT), activation (ACT) and hostility (HOST) reveals both samples as severely afflicted by the psychosis. Also with the scale for negative symptoms (SANS), positive symptoms (SAPS) and global achievement (GAS; see Andreasen 1987) male and female patients seem equally affected by the disease. Up to now 17 of these patients and 35 controls have been examined by MRI. The latter are not identical with the above-mentioned control sample. Therefore an intragroup comparison can be only done with the patients sample.

As can be seen in Fig. 1, music 1 - the repetitive arpeggio - elicits quite different delta-power changes in controls and patients. Controls show a general decrease over the whole skull with marked effects mainly over frontocentral and temporal regions in both hemispheres. In comparison, delta power was augmented in the patients over all electrode places besides both temporal positions T1 and T2. Nevertheless a significant difference with the controls also resulted here, whereas in most other electrode places the difference was even highly significant. This means a nonreactivity or even a diffuse delta activation in schizophrenics with no local preference.

Figure 1 shows the variation in alpha power during music 1 stimulation. In controls and patients an alpha-wave decrease is seen in frontocentral and centroparietal electrodes which tends to be smaller in patients than in controls. Over right posterior electrodes this different reaction is significant. In the region of the apex of the right temporal lobe a slight increase in alpha power was registered in the patients.

Figure 1 contains the delta power results with music 2 stimulation, the Mozart quartet. In controls, there is a general decrease comparable to that with music 1, which is also maximum over frontocentral and frontotemporal regions. In contrast to music 1, delta activity decreases in patients in these electrodes during music 2. But compared with controls, they decrease to a lesser extent. With preference to right temporoparietal and parietocentral positions even an increase is seen, which means there is a significant difference over right temporal cortex and also left posterior cortical areas compared with controls. As is seen in Fig. 1d, alpha power decreases in controls with music 2 as compared to music 1. But with Mozart also a left temporal decrease is obvious, indicating involvement of those cortical areas. In contrast, alpha waves are increased in both temporal areas as well as in frontopolar electrodes. The frontal and the right temporal augmentation distinguishes both samples significantly.

From 19 MRI variables defining cerebral atrophy (see Vogel 1986) results from the following nine differentiated schizophrenics from controls: Frontal horn index 1 and 2 (FHI1/2), the Huckman number (HN), third ventricle (V3), left ventricle temporal horn width (THL, THMX), width of right cisterna fossae lateralis cerebri (fissura Sylvii, CFLSR, CFLMX), and visible sulci of vermis cerebellaris (NSCL). No differences were found in cella media index 1 and 2, ventricle index V1/2, width of fissura longitudinalis (FLC), right temporal ventricle, left fissura Sylvii and ventricle/fissura Sylvii index.

Since in 17 patients who had been examined with MRI EEG mapping was also done, correlation coefficients between morphologic and functional data could be defined. As is shown in Table 2, correlation coefficients were calculated for 21 electrode positions, in which delta power within the reference state was significantly higher in schizophrenics than in controls. *t*-Testing reveals some meaningful negative correlations between the degree of atrophy and delta production in the relaxed state. Only schizophrenic patients with a narrow third ventricle produce delta-waves over the whole cortex. An atrophic right fissura Sylvii is negatively related to delta production mainly in centroparietal and temporoparietal areas. This also holds for

Table 2 Coefficients of the correlation between EEG mapping data in resting state and MRI-variables delta frequency

	FHI2	V3	THL	THMX	CFLMX	NSCL
FP2		-.5376				-.3192
FP1		-.4774				
F4		-.5474				
FZ		-.5451		-.334	-.3983	
F3		-.4536				
FC6		-.3614				
FC2		-.5255		-.3107	-.4261	.3344
FC1		-.4338			-.3641	.3373
C4		-.5247	-.3197	-.3096	-.3531	.3261
CZ		-.4698	-.3284	-.3460	-.4650	.3415
C3		-.4312			-.4187	.4559
CP6		-.4408	-.3692	-.3374	-.3045	.3063
CP2		-.5992		-.3236	-.5624	.5105
CP1		-.4876	-.3054	-.3291	-.5281	.4947
CP5		-.3100			-.4332	.3719
P4		-.4812	-.3174	-.3220	-.4233	.5044
PZ		-.4441		-.3228	-.5459	.4753
P3		-.3653		-.3012	-.4780	.4884
O2						
OZ			-.3335			
O1	.3140	-.3419	-.4955	-.4778	-.4049	

Only correlation coefficients above 0.3 are listed. MRI variables see text. Electrode positions according to the 10-20 system (see Fig. 1).

an enlarged left ventricle. Otherwise cerebellar atrophy seems to correlate positively with delta power.

When listening to music 1, a quite different function of the atrophic brain seems to emerge in the relation of atrophy compared to the resting state, as is shown in Table 3. Frontal atrophy seems to reduce delta production considerably, independent of the hemisphere involved. Otherwise a positive correlation results between Huckman number, which represents inner frontal atrophic processes, and delta production in frontocentral regions. Atrophy of the third ventricle has only some tendencies of negative correlation in central parts. Otherwise a positive correlation between left temporal ventricle atrophy and delta production in frontal areas is obvious, as is the correlation between cerebellar atrophy and occipital delta power. Right fissura Sylvii atrophy seems to increase delta-wave production within this area (F8, T2), whereas centroparietal electrodes show a negative correlation.

Table 3 Coefficients of the corrlation between EEG mapping data under music stimulation and MRI variables, delta frequency band

	FHI1	FHI2	HN	V3	THL	THMX	CFLMX	NSCL
FP2	-.3215	-.4537			.4329	.3274		
FP1	-.3764	-.4927			.5151	.3964		
F8		-.3416			.4074	.3079	.3282	
F4	-.3432	-.4254	.3629		.3698	.3015		
FZ	-.3712	-.4380	.3247		.3321			
F3	-.3890	-.3994	.3360		.3114			
F7		-.3844	.3026		.4145	.3151		
F2							.3184	
FC6	-.3825	-.4541	.3962		.3270			
FC2	-.3430	-.3719	.3025					
FC1	-.3366	-.3473	.3242					
FC5		-.3180						
T1								
T4	-.4502	-.4165	.5072					
C4								
CZ				-.3748			-.3591	
C3				-.3022				
T3	-.3307	-.3298	.3220					
CP6	-.3055							
CP2				-.3988			-.3224	.4251
CP1				-.3480			-.3773	.3519
CP5								
T6								
P4				-.3315			-.3243	.5165
PZ							-.3146	.4915
P3								.3308
T5								
O2								.3712
OZ					-.3688			
O1								.3161

Only correlation coefficients above 0.3 are listed. MRI variables see text. Electrode positions according to the 10-20 system (see Fig. 1).

Discussion

Data presented here are the preliminary results of an extensive interdisciplinary study aiming at correlating and validating different methods of morphologic and functional examinations of the cerebrum in mental patients and controls. Up to now only data from controls and schizophrenic patients can be presented. Nevertheless some results of previous studies are already confirmed, which can be taken as justification for a discussion even at this early stage. This does, of course, limit reliability.

Concerning the sample size, EEG mapping data are representative both in patients and controls also with respect to gender. Nevertheless a correction for age in the female and male sample has not been done up to now since a truly comparable sample would attain the limits of meaningful statistics. For the same reasons the MRI-sample could not be split according to sex. Therefore mapping data were also calculated irrespective of sex, although in former studies (Petsche et al. 1988; Steinberg et al. 1992) a considerable influence of gender could be shown. The explorative nature of the examination will allow for an early discussion, since a clear influence of patient age (in their twenties and thirties) was never shown. With comparable arguments a Bonferoni/Holm adjustment of the data was not exerted nor was one-tailed *t*-testing applied. A strong influence of musical achievement upon EEG data within this sample will be low, since no musician or skilled amateur took part in the study.

Previous results with musical stimuli in patients and controls could be confirmed. Acute schizophrenic patients show significantly higher delta activity over the whole cortex than controls which even increases under a repetitive, simply structured harmonic - rhythmic stimulus. In contrast, delta activity decreased in controls mainly in frontotemporal areas of both hemispheres. Taking a "real" musical stimulus, Mozarts quartet, for example, the difference in delta reaction remains between patients and controls, although in both groups some differences in comparison with the monotonous stimulus appears. In the patients delta-waves decrease in frontal regions, too, but over both temporal electrode positions, which are next to the primary auditory cortex, a significant delta increase prevails. Also within the alpha frequency band a pronounced difference between both samples shows up. Whereas they decrease in controls over both temporoparietal regions, in schizophrenics there is no reaction or a slight increase, which under Mozart stimulus is significant. Here a decrease in left temporoparietal regions also occurred in controls.

Regarding previous mapping studies with musical stimulation in healthy controls (for review see Steinberg et al. 1992), a decrease in delta power compared to resting conditions could be confirmed. With both monotonous and literate music stimulation slow brain activity diminished mainly frontotemporally with no clear hemispheric preference. Otherwise, a harmonic rumba rhythm decreased alpha power mainly in the right hemisphere, whereas Mozart also involves the left hemisphere. This seems somewhat different from dichotic listening experiments, which assume a representation of harmony and wholistic listening preferentially on the right site, whereas rhythmicity and sequential aspects are confined to the left hemisphere. In former studies the clear difference of alpha and delta power variation in patients had disappeared after 2 weeks of psychopharmacologic treatment, which resulted in an at least partial remission. Therefore, the difference between patients and controls shown here might well be due to the acute disease. Whether former results can be confirmed remains to be proven with follow-up data.

The correlation between function and morphology, resulting from calculating the coefficients between mapping and MRI data, gives some promising hints. Although most of these coefficients can only be seen as a statistic trend, it is surprising that the low resolving power of MRI and EEG methods show a correlation of the data at all. Taking these results as an indication for a true correlation, some hypothesis concerning the pathways involved can be discussed. General atrophy within brainstem structures adjacent to the third ventricle prevents delta-waves production in schizophrenics in resting conditions, which also seems to hold for some other atrophic processes, as for example, in the left ventricle and right fissura Sylvii. Why a supposed cerebellar atrophy seems to sustain delta-waves production cannot be answered. Otherwise, when applying a simple monotone musical stimulus, the involvement of frontal areas in music processing seems evident by the clear dependence of delta production of intact morphologic structures. But a central atrophy of structures adjacent frontal ventricular parts and also left ventricular horn, as is quantified by the Huckman number and THL, seems to foster delta production in musical stimulation. Which cortical layers or which central pathways are involved in delta production cannot be answered by MRI atrophy data. That right temporal areas adjacent to Heschl´s gyrus are involved in delta production can possibly be taken as an argument for involvement of primary auditory cortex. Taking also into consideration an involvement of primary cortical areas in hallucinatory processes, as evidenced with SPECT data in a casual report (Erkwoh et al. 1992), general involvement of these cortical areas in the schizophrenic process attains some probability.

The results presented do not contradict the hypothesis of no reaction of type II schizophrenics seen in glucose utilization or augmentation of cerebral circulation, as was demonstrated both with musical and motor stimulation. Since these results were from patients not treated pharmacologically or even when schizophrenia shows up for the first time, morphologic alterations and functional differences seem to exist before admission; such results could also be shown by others. Therefore musical stimulation is an adequate means for studying cognitive processes in controls and mental patients.

References

Andreasen NC (1987) The diagnosis of schizophrenia. Schizophr Bull 13: 9-22

Beckmann H, Gattaz WF, Jakob H (1987) Biochemical and neuropathological indices for the aetiology of schizophrenia. In: Häfner H, Gattaz WF, Janzarik W (eds) Search for the causes of schizophrenia. Springer, Berlin Heidelberg New York,pp 241-249

Crow TJ (1985) The two syndrome concept. Origins and current status. Schizophr Bull 11:471-486
Erkwoh R, Ebel H, Kachel F, Reiche W, Ringelstein EB, Zimmermann J, Büll U, Saß H (1992) Musical and verbal halluzinations correlated to electroencephalographic and PET-findings. Case report. Nervenarzt 63:169-174
Günther W, Streck P, Steinberg R, Günther R, Raith L, Backmund M (1989) Psychomotor disturbances in psychiatric patients as a possible basis for new attempts at differential diagnosis and therapy. IV. Brain dysfunction during motor activation measured by EEG-mapping. Eur Arch Psychiatr Neurol Sci 239:194-209
Günther W, Steinberg R, Streck P, Banquet J-P, Bscheid I, Raith L, Riedel R, Klages U, Stiltz I (1991)Brain dysfunction in psychiatric patients during music perception measured by EEG-mapping: Relation to motor dysfunction and influence of neuroleptics. Eur Neuropsychopharmacol 1:143-155
Overall JE, Gorham DR (1962) The brief psychiatric rating scale. Psychol Rep 10:799-812
Petsche H, Lindner K, Rappelsberger P (1988) The EEG: an adequate method to concretize brain activity elicited by music. Music Percept 6:133-160
Raith L, Steinberg R, Reitschuster S (1989) Amusia and mental diseases. Musik Tanz Kunstther 2:121-126
Steinberg R, Raith L, Eben E, Roßnagl E (1985) Music-psychopathology. III. Musical expression and psychiatric disease. Psychopathology 18:274-285
Steinberg R, Raith L, Günther W (1989) Hearing and perception in different mental states. In: Nielzén S, Olsson O (eds) Structure and perception of electroacoustic sound and music. Amsterdam: Excerpta Medica Amsterdam,pp 106-116
Steinberg R, Günther W, Stiltz I, Raith L, Rondot P (1992) EEG-mapping during music stimulation. Psychomusicology 11:154-169
Vogel H (1986) Maße in der Sonographie und Computertomographie. Ecomed, Landsberg

Subject Index

GPSR Compliance

The European Union's (EU) General Product Safety Regulation (GPSR) is a set of rules that requires consumer products to be safe and our obligations to ensure this.

If you have any concerns about our products, you can contact us on ProductSafety@springernature.com

In case Publisher is established outside the EU, the EU authorized representative is:

Springer Nature Customer Service Center GmbH
Europaplatz 3
69115 Heidelberg, Germany

Batch number: 10844669

Printed by Printforce, the Netherlands